THE DEADLY BALANCE

Some other titles in the Bloomsbury Sigma series:

THE DEADLY BALANCE

*Predators and People in a
Crowded World*

Adam Hart

BLOOMSBURY SIGMA
LONDON · OXFORD · NEW YORK · NEW DELHI · SYDNEY

BLOOMSBURY SIGMA
Bloomsbury Publishing Plc
50 Bedford Square, London, WC1B 3DP, UK
29 Earlsfort Terrace, Dublin 2, Ireland

BLOOMSBURY, BLOOMSBURY SIGMA and the Bloomsbury Sigma logo are
trademarks of Bloomsbury Publishing Plc

First published in the United Kingdom in 2023

ISBN: HB: 978-1-4729-8536-1; eBook: 978-1-4729-8532-3

2 4 6 8 10 9 7 5 3 1

Typeset by Deanta Global Publishing Services, Chennai, India
Printed and bound in Great Britain by CPI Group (UK) Ltd, Croydon CR0 4YY

Bloomsbury Sigma, Book Seventy-Seven

MIX
Paper | Supporting
responsible forestry
FSC® C171272

To find out more about our authors and books visit www.bloomsbury.com
and sign up for our newsletters

Contents

Dedicated to all those who live with dangerous animals across the world and strive to keep them safe from us – and us from them.

CHAPTER ONE

Introduction

In August 2015, I was tracking lions in the South African bush. An old male lion, known by local tourist guides and researchers as Cecil, had been killed in Zimbabwe about six weeks before by the trophy-hunting American dentist Walter Palmer. I am sure you remember it. There was a huge furore across the media and as a consequence trophy hunting – and lions – were in the public consciousness like never before. I was in South Africa with the BBC Radio Science Unit to try to uncover some of the facts and figures around hunting generally, and lion hunting in particular.[1]

Walking around Pilanesberg National Park on a sunny afternoon with an armed ranger, a hugely knowledgeable tracker (unarmed) and a producer (armed only with recording equipment), it didn't take us long to find signs of lions. There is a healthy and growing population of these big cats within the park and I have seen them there many times. There is a confident power and fluidity about lions that help to make them the species most safari tourists want to see. They really are the king of beasts, but – as the signs say – you'd be wise to 'stay in your vehicle'. The prospect of being on foot with a lion or – as is likely in a species that tends to live in groups – multiple lions, was making me more than usually vigilant that afternoon. Added to that was the growing realisation that lion viewing felt very different without the protection of metal and glass. The lion tracks we were following had a very familiar cat-paw shape, but they were absolutely huge. Just about managing to bracket my sunglasses around a particularly clear print, I hovered over it to take a photograph. Despite a cooling breeze and the gentle pace of our tracking, a droplet of sweat fell from my forehead on to the footprint. Our casual chit-chat had dried up. The mood had changed.

One thing you learn early in tracking is the mantra 'sun-track-tracker'. By positioning yourself correctly in relation to the sun, the shadows that form on the ground from the ridges and depressions of animal tracks can reveal signs that were invisible from another vantage point. Tiny drag marks, small pad indentations and the subtlest of indications can ping out if you get your viewing angle just right. The tracks we now saw moving away from us had no need for careful 'sun-track-tracker' positioning. They were big enough – and fresh enough – for us to have followed in starlight if we had to. Picking a tiny hair off the mark left by a giant rear paw – and disturbing a sharply defined ridge of sandy soil that the wind had yet to smooth out – pointed to only one conclusion. This lion, a big male, was close.

Looking around us, the ground had changed. As can happen when you are focused on a task, you don't always notice gradual changes creeping up on you. For the past few hundred metres we had been heading steadily into a mishmash of narrow, deeply cut drainage channels. The sides rose up 3 metres and more from flat ground covered in sand deposited by seasonal floods. An ever-narrowing labyrinth of eroded sediments drew us in, and up ahead lay a section where forward visibility was reduced to just a few metres in the twists and turns. Suddenly, the goosebumps came. They are coming again now as I type. There was the absolute, cold realisation that we were no longer the 'hunters'. We were in danger of becoming prey – and it is a feeling I will never forget.

We never did see the lion we tracked. Looking at each other in that gully, we all realised that discretion was the better part of valour. As much as a lion encounter would be excellent for the programme, this particular lion, in this environment, was best left alone and we made a determined retreat. Later that evening our vehicle headlights illuminated a fresh zebra kill surrounded by a pride of blood-soaked lions enjoying their dinner. At least, we thought, that wasn't us entering the 'circle of life'.

The possibility of being hunted down, killed by teeth and claws, and ending up in the belly of a predator is not something of great concern to many of us in the modern world. It may be a deep-seated fear, a nightmare even, but it is neither a lived reality nor a faint possibility. Living in rural Gloucestershire, for example, the most dangerous animal I am ever likely to face while out for a walk is an aggressive dog. Lest we make too light of that, we should recognise that dogs can be a very real threat. In the UK people occasionally die as a result of dog attack. In evidence submitted to DEFRA in May 2018, 31 deaths from dog attacks have been recorded in the UK since 2005, including 16 children and 15 adults.[2] At the time of writing, the most recent fatal attack involved a two-year-old boy.[3] So, being killed by a dog, a predator (albeit one that has been heavily genetically modified over the course of centuries), is still a possibility, even if you live somewhere not known for hazardous wildlife. But fatal dog attacks, truly awful and avoidable though they are, are rarely predatory interactions. People who are killed in dog attacks are usually not eaten and the 'predators' are not motivated by hunger (but see Chapter Nine for some exceptions).

Such is the relative harmlessness of the fauna of many developed world nations that most of us have the luxury of viewing the natural world as a place of wonder, solace and calm. The notion that serious harm or even death could result from an interaction with wildlife is far from our minds. In some parts of the world, though, there is a huge number of potentially lethal animals that people have to live with on a daily, and nightly, basis.

Broadly speaking, lethal animals occur in two types: those that kill with biochemistry and those that kill with physics. Biochemical killers – like venomous snakes, scorpions, spiders, wasps and bees – kill with molecules that have evolved for subduing prey, for defence or both. Snake venoms, for example, often attack either the nervous system (neurotoxins) or various tissues including components of blood (hemotoxins). When snakes strike prey, they do not want it to fight back, or to travel far before it expires; a snake could easily lose its dinner to an opportunistic bird or mammal if it had to track it too far. Consequently, selection has tended to favour highly potent venoms that get the job done quickly. The result is a very rapid death for a rodent or small mammal, and a major problem for a larger animal that may end up on the wrong end of a defensive snake strike. Some snake venoms can be medically extremely serious. Highly venomous species like the fer-de-lance of Central and South America, or the black mamba of southern Africa, are greatly feared by local people for good reason. Without antivenom, a black mamba bite is very often fatal. However, if you are unlucky enough to get bitten by a black mamba I can guarantee you one thing: you won't be eaten. Black mambas and other venomous snakes bite humans in self-defence, striking out of fear rather than hunger. Likewise, although honeybees and wasps kill dozens of people every year in the United States and somewhere between two and nine people in the UK, none of these are predatory killings. Stings have evolved that hurt and deter, and the few deaths that do occur usually result from the victim's severe allergic reaction to the venom.

Other animals kill us with force, often in conjunction with hard 'accessories' like teeth and claws that concentrate the force the animal can apply into a small surface area. A force delivered over a smaller area results in more pressure at the surface. This is high-school physics in action. Teeth and claws can penetrate skin, muscle and even bone causing serious and sometimes catastrophic injuries. Google 'bear attack victims' if you want to see the sorts of injuries I am talking about, but I don't suggest you do so if you are at all squeamish.

As is the case with the biochemical killers, most attacks on humans from animals with teeth and claws are not predatory but defensive. Animals usually choose to avoid us rather than risk a confrontation and will only engage with us as a last resort. If we want to consider only those animals that have predation as the sole motivation for interacting with us then our pool of candidate species shrinks considerably. There are only so many animals that combine the right qualities to even begin to consider us as prey. For a start, a potential predator needs to be relatively large. We are a fair-sized mammal, with reasonable strength and a decent all-round sensory capability. This suggests that a potential predator needs to be at least in the same ballpark physically as us to have much of a chance. That being said, the fact that a fully grown adult can be killed by a dog shows that, on our own, we are potentially vulnerable even to smaller species if they are able to overwhelm us and get us to the ground. We are sociable, though, and smart. We are often in groups and our strength in numbers – as we will see – provides considerable protection against predators. More protection is afforded by our unparalleled ability to use the environment to our advantage, fashioning weapons that our opposable thumbs, upright stance, strong and mobile upper limbs, and relatively large brains let us wield to great effect. So, while there are plenty of predatory species out there – and however fearsome honey badgers or wolverines may appear – most aren't looking at us as prey.[4] To be a potential predator of humans, we need to think bigger (while keeping in the back of our minds the fact that younger humans, and especially infants, are a very different proposition for potential predators).

Large carnivores like the big cats, some of the canids (the 'dogs', notably wolves), hyenas and bears fall into the right size range, as do the larger crocodilians, the bigger varanid lizards (most obviously the Komodo dragon) and a few species of snakes. Rivers and lakes can be home to some very large fish species, a number of which have been implicated in predatory attacks on people, although size may not always be a factor in the freshwater environment, as the many tales of piranha attacks suggest. Strength in numbers could work on land too. Tales of army ants, massing in their hundreds of thousands and pouring over a hapless victim as an unstoppable insect tide are the stock-in-trade of 'jungle thrillers', but – as we will see – the reality doesn't always match the imagined terror. In the oceans there are many potential predators to choose from, not least because there is a multitude of larger predatory shark species and a good diversity of other suitably sized predatory fish. A number of whale species could also be candidate predators, our relative vulnerability in the water adding perhaps to our attraction as an easy meal. Even seals might get in on the act. There is at least one incident of a leopard seal attacking and killing a human – marine biologist Kirsty Brown – working in the Antarctic in 2003. A subsequent study of leopard seal interactions concluded that predation was the motivation, adding another species to our growing list.[5]

Faced with a large and hungry predator, you might think that we are pretty defenceless. Certainly, in the water we are at a disadvantage, but even on land there seems little we can do against a predator like one of the big cats. A fully grown male African lion comfortably weighs more than 180kg (397lb), and has claws that are strong, sharp and long (around 3.5cm long) and a mouthful of 30 teeth evolved for meat-eating and flesh-tearing. At 83kg (183lb) wet, with trimmed nails and teeth more suited to nibbling than tearing, I don't feel that I stack up well regardless of how effectively I can use my environment to improvise weapons. There is a reason why the ranger I was walking around with at the start of the chapter was carrying a rifle. There are, however, stories of people fending off and even killing predators with little more than cutlery.

The most celebrated of these encounters involved the first game ranger of Kruger National Park in South Africa. A tough, knowledgeable and determined man, Harry Wolhuter served the park for 44 years from 1902. Patrolling on horseback in 1903, early in his tenure, Wolhuter was charged by a lion. Knocked off his mount, he watched the horse run off, pursued by the lion, which was in turn pursued by his dog. With the sun sinking rapidly, his problems had only just begun. A second lion was now stalking him from behind, a fact he noticed too late to avoid an attack. The lion sprang on to him, biting him deeply in the shoulder. Wolhuter was then dragged towards a spot where the lion could finish the job and make a meal of the unlucky game ranger. During this drag – a predatory behaviour we will see again in Chapters Two, Three and Seven – Wolhuter remembered his knife, safely sheathed on his belt. Pulling out what was basically a decent-sized kitchen knife,[5] he stabbed the lion three times, twice in the torso and once in the neck. The lion limped away and Wolhuter climbed a tree, bleeding heavily. It was at that point that the first lion, last seen in pursuit of Wolhuter's horse, reappeared. Wolhuter must have been feeling he was having rather a trying day but, long story short – as they are fond of saying in South Africa – his men eventually came and rescued him. His wounds were bad and became so infected that his doctors feared the worst. Wolhuter pulled through, though, and went on to live a full and exciting life, dying in 1964. This is in marked contrast to the lion he stabbed, whose skin is on show, together with Wolhuter's knife, at the Stevenson-Hamilton Memorial Library in Skukuza, Kruger National Park.[6]

Despite our brain power, ability to use weapons and a resilience and determination amply demonstrated by Wolhuter, even my casual survey of the animal kingdom revealed a great many species that could easily consider us as prey. Famous historical accounts of 'man-eaters' – from the renowned maneless man-eating lions of Tsavo (Chapter Two) to Jim Corbett's tigers (Chapter Three) and leopards (Chapter Seven) – illustrate that we can become prey, but also reveal much about how the world has changed over the past century or so both for humans and their potential predators. Looking even further back, our folklore and legends can tell us much about our relationship with predatory animals in the deeper past, when our interactions with nature were more intimate and immediate. To a great extent, much of our impression of 'man-eaters' is rooted in this past, but occasional – and often lurid – contemporary news stories confirm that such things also happen today. It will become a common theme through the following chapter that the reports which surface only rarely in mainstream Western media come nowhere close to revealing the true extent of human depredation in parts of the world. Make no mistake, for some people today the fear of becoming prey is very real indeed. Also, this paragraph marks the last time I will use the term 'man-eater' unless in a direct quote. It implies that the victims of such attacks are always men and that is very far from the case.

I have already emphasised the difference between animals that attack us (even with lethal consequences) and animals that hunt us for food, but there is another more subtle distinction that needs to be made: that between 'animals that eat us' and 'animals that make us prey'. If a human body is left for any period of time then blow flies and flesh flies rapidly move in and lay eggs on it. These eggs hatch within days to form squirming masses of maggots that quickly convert human flesh into more flies. Dermestid beetles (commonly known as 'hide beetles' or 'skin beetles') can change the desiccating hide of a human into more beetles just as surely as they can that of a dead deer, sheep or rat. Insects have to compete for flesh with the larger members of the scavenger community. Big carcass breakers like hyenas (Chapter Six) quickly move in on any freshly dead animals, filling their bellies and providing opportunities for scavengers like vultures, crows and jackals (Chapter Nine) to enter and more effectively pick clean a carcass. Even herbivores can join in the feast. I once saw a horse munching away on a dead wood pigeon, and the loudly unpleasant sound of a giraffe chewing on the vertebrae of a wildebeest is one I find hard to forget. There is no reason at all to think that humans are immune to all of this feasting, somehow granted by nature a special dignity in death by dint of our intelligence or achievements. We are, ultimately, meat and bones, and left out in the open are just as attractive to scavengers as any other animal. However fascinating and essential the scavengers and decomposers are, it is the predators that make us prey and it is therefore predators that are the focus of this book. Except sharks. There are far too many books and documentaries out there discussing shark attacks, so I'm going to stick to land animals.

I have a bigger ambition for this book that extends beyond detailing the many ways in which humans can end up as prey. Predators are essential to properly functioning ecosystems, but our fear of them killing us and our livestock has been a major factor in their historical and contemporary persecution. Despite these fears, predators exert a powerful hold over many us. They are romanticised, anthropomorphised (Disney's *The Lion King* being a fine example) and seen as spirits of the wilderness (wolves and bears especially). And despite the dangers, or perhaps because of them, many of us with the luxury of a safe, predator-free life hanker to see them in the wild, sometimes to witness them make a kill and to experience their power as close to 'first hand' as we can. By examining the conflicts that develop between humans and predators, the complexity of our relationships with predators and the ways in which we can enhance coexistence, I want to find some hope that the world of the future will be a world where predators can live without the fear of persecution and humans can avoid becoming prey.

CHAPTER TWO

Lions

I would like you to try to imagine that you are a rural Tanzanian villager. This isn't going to be easy, but it's important to try. For the purposes of this imagineering, you live in the Lindi region, which is next to the coast in the south-east corner of Tanzania. Feel free to take a look on Google Earth if it helps you to picture the place. If you do, you'll notice one thing pretty quickly: Tanzania is a big country. At just over 945,000km² (365,000 square miles) it is larger than Spain, Portugal, Italy, Belgium (that geographical staple of area calculations) and most of Wales combined. It shows just how large African countries are that – while being larger than substantial chunks of Europe combined – Tanzania still only manages to rank a paltry 13th in the table of largest countries in Africa. In fact, with 54 recognised countries across Africa, Tanzania only just squeaks into the top quarter.

The Lindi region, where you live, is 'just' 66,000km² (25,000 square miles) or a little over a 'pair of Belgiums'. That said, if the Lindi region were an African country, rather than just one of the 31 regions of Tanzania, it would be the 41st largest, just between Sierra Leone and Togo. I guess the point I am trying to make here is that when we talk of 'Africa', it is rare for many people who haven't had the opportunity to spend time there to have much of an idea of just how vast some of the areas concerned are. This sense of scale is important, because when we are thinking about predators and human – predator interactions in many places across the world, geographical scale is going to prove to be a vital consideration. Among other things, scale influences remoteness, relative development, predator abundance, biodiversity and human population density. Of the 31 regions, or *mikoa*, that make up Tanzania, your region of Lindi has one of the lowest population densities. Tanzania overall, though, has an expanding population. Indeed, East Africa (which includes Tanzania, Kenya, Uganda and Ethiopia) is one of the fastest growing regions in the world. The extent of this expansion can be gauged simply by looking at population trends over the past 20 years. At the start of the century there were 35 million people in Tanzania, which by 2018 had risen to 56 million. This an increase of 21 million, or 60 per cent, in 18 years. In 1988, there were just 23 million people in Tanzania, which means the population has almost tripled in 30 years. Population rarely expands evenly across countries, especially larger countries, and in Tanzania most population growth has been concentrated in the north. Together with a general trend of urbanisation over the past 50 years, expanding northern populations mean that the southern regions, like Lindi, still remain relatively undeveloped, at least for now.

The Lindi region has just under 900,000 people and is split up into five districts. The district you call home is Ruangwa. By this stage of geographical division, we are getting into area sizes that are more familiar and comfortable. Ruangwa is the smallest district in the

Lindi region, with an area of around 2,500km^2 (965 square miles). This is more or less the size of the English county of Dorset. You and your family live, grow crops and graze livestock here.

One sign of the undeveloped nature of the Ruangwa district is the general lack of paved roads. One estimate puts this as low as 5km (3 miles). These days we have the advantage of being able to check out such claims without leaving our homes, thanks to Google Earth. Heading down to the Ruangwa district, the initial zoomed-out images make the whole region look like wilderness. As in many other parts of Tanzania – and indeed the world more widely – what looks like wilderness very often isn't. Zooming in closer reveals the distinctive signs of a human-dominated – or at least human-affected – landscape. Once we get even closer to the ground, a mosaic of regularly sized, more or less straight-edged cleared areas comes into view. It can be hard to get your eye in at first, but once you have, cultivated areas can easily be distinguished, spread throughout much of the district. Zooming in yet further reveals huts and other signs of habitation, as well as tracks across the ground made by wildlife, people and vehicles. Flying virtually around the region, the '5km of paved roads' claim starts to look like a realistic estimate, but despite the lack of infrastructure development it is hard to escape the signs of human presence. Zooming into a patch of woodland reveals the regular pattern of a plantation, while a closer look at a patch of scrub shows more than 10 huts within cleared areas of bushes and trees. Some parts of the district are less amenable to cultivation and there it is harder to find signs of human presence. Thick bands of dark vegetation, for example, mark out drainage lines and seasonal rivers, while thicker stands of trees seem relatively resistant to human impacts, at least for now. So, Ruangwa may not be a heavily developed area, and population density is relatively very low, but it would be a big mistake to think that this is truly 'wild' country. As we will see, it could end up being an even bigger mistake to think of it as 'tamed'.

So, there you are, living a rural existence in southern Tanzania. It is a lifestyle that wouldn't be so different in a great many other countries throughout the world. Millions, perhaps billions of people live some distance from cities and advanced infrastructure, but nonetheless they are people who are very much part of the modern world. Just like you, these are people who need food and shelter, water and power, people with hopes and aspirations, people who love their children and tolerate their in-laws. This is important to keep in mind because quite soon we are going to have to think about people being eaten, and this will inevitably involve numbers and statistics. When those numbers grow large, the humanity that contributes to them becomes obscured and forgotten. But we must never lose sight of the fact that behind these numbers are real people. They might live far away, and they might have a different culture and lifestyle, but in every way that really matters they are just like you and me.

Speaking of you, it just so happens that when we join you down in Tanzania it is mid-April, and the heaviest rains are falling. This might make moving around more difficult at times, and even hazardous as dry river beds fill with rapid floods, but your crops of cassava, maize and sorghum have grown well and you are spending long days in the fields harvesting whatever is ready. It isn't just humans that find your crops attractive; bush pigs are present in good numbers around your fields. At night, these large, intelligent and powerful relatives of the domestic pig roam in groups of perhaps a dozen or more individuals. To help to protect your crops from them you are sleeping in a makeshift hut. A decent enough construction of branches and twigs, your hut is raised on a platform less than 2m (6ft 7in) above the ground and provides reasonable shelter from the elements.[1] Bush pigs are a great example of the label 'pest' being highly subjective. They might be the last thing you want to be around if you are a Tanzanian farmer, yet they are a species some wildlife tourists are desperate to see. It is all about perspective – and from your current perspective, a sleeping platform above the ground, they are more than just a pest. You see, the problem with bush pigs is that, as well as eating your crops, they attract lions – and Tanzania has more lions than any other country.

Lions past and present
Some of us think of lions these days as a 'safari' species, prowling the endless plains of Africa. Their modern-day range backs up this impression. Most lions are now found in southern and eastern Africa, and they are a much sought-after species by tourists on safari to destinations like Kruger National Park in South Africa or Hwange National Park in Zimbabwe. The modern lion range, though, is a hugely contracted fragment of

the species' historical range. Lions were once dispersed far more widely across Africa, as well as parts of southern Europe and Asia. For example, lions were once commonplace in Greece, although they seem to have disappeared from there at least 1,500 years ago. Lions were also present throughout the southern and eastern Caucasus until around 1,000 years ago. Now, wild lions outside of Africa are limited to the Gir National Park and surrounding area in the western India state of Gujarat. These Indian lions are also but a tiny remnant of a Central Asian and Indian distribution that included Saudi Arabia, Iran, Iraq and much of the northern swathe of India. We are fairly and squarely to blame for this historical range contraction and for the modern-day threats to lions. Although the precise reasons behind any decline can be complex, the overarching themes are ones we will return to frequently in this book: predators are persecuted because they are dangerous to us and our livestock; we reduce their habitat through encroachment; and we deplete their prey by hunting.

Within Africa, lion range has greatly contracted over the past century or so. This decline in geographical range goes hand-in-hand with a huge decline in overall population. Lions are now either absent or extremely rare in most of North and West Africa. They have been extirpated (made locally extinct) from Mauritania, Gambia, Togo, Djibouti, Lesotho and very probably Sierra Leone (a country with the Italian word for 'lion' in its name) and Eritrea. It is also likely that lions are no longer present in Ghana, Guinea, Guinea Bissau, Ivory Coast and Mali. Tantalising camera-trap images suggest that lions may be present in a few countries formally written

off but, from a historical range that included most of
Africa except the Sahara Desert and the rainforests of the
Congo River Basin, lion range has contracted to perhaps
a quarter of what it once was.

In terms of overall population, estimates vary. Lions
are difficult to survey in many parts of their range and –
despite the global attention paid to lion conservation –
funding for on-the-ground research and accurate
counts is hard to find. Even when groups and individuals
are funded to undertake such research, population
estimates are just that – estimates. Keith Somerville
provides a good overview of lion populations in his
book, *Humans and Lions: Conflict, Conservation and
Coexistence*, and he concludes that the general opinion
of conservation scientists working on lions is that the
population is somewhere between 20,000 and 39,000.
Amy Dickman, one of the world's top lion researchers
and the leader of the Ruaha Carnivore Project in Central
Tanzania, suggests a likely number somewhere just
under 24,000, while others stand behind a figure of
20,000. The simple answer is that we don't know for
sure, but I've yet to see a credible estimate that doesn't
fall somewhere between 20,000 and 25,000.

Overall population estimates are useful, of course, but
only up to a point. Current population does not give us
any idea of how the population has changed over time.
Without the context provided by historical population
data we can have little hope of gaining insight into
potential future population trends. Once we factor in
historical estimates, the situation with lions looks even
more dire. We have enough of a problem gaining reliable
estimates of the population now – when lions are

greatly restricted in terms of range and we have modern survey methods – so historical population estimates have even more error associated with them. With that in mind, it is sobering to learn that best-guess estimates suggest that lions may have numbered 400,000 in 1950 and perhaps as many as 100,000 in the mid-1990s. I wouldn't want to bet my house on whether those figures are even within 10 per cent of the actual number, but regardless of the detail the bigger picture is horribly clear.

What overall population figures also mask is local variation. To sidestep slightly into some fundamental ecology, all species have an 'overall population'. However, unless a species is so limited in its range that all members live within 'mixing range' of each other, such headline figures are less useful than knowing about how that overall population is dispersed through the species' total range. In reality, most species exist as different, more-or-less separate populations within the overall range. Exceptions to this are species whose range is limited to a single location, such as those endemic to small islands. In most species, though, individuals from different populations within the total range may have only limited contact with each other, if any contact at all. There may be some immigration and emigration from different populations and this can be important to limit the harmful effects of inbreeding, especially when population size is low. Despite considerable dispersal ability, however, a lion living in Ruaha National Park in Tanzania will never meet a lion prowling Pilanesberg National Park in South Africa. Individual lions from each population would have no problem breeding with each other – they are undeniably the same species – but they will never breed simply because they will never meet.

To conserve species, it is vital to understand population ecology because it lets us understand the specific threats faced by different populations and then tailor our conservation approaches accordingly. So, while lions are in decline overall – and their range is contracting – if we examine individual populations of lions across their range we see a more nuanced pattern. In some areas, lion populations are declining calamitously, with populations in West and Central Africa of particular concern. In parts of southern Africa, however, we see some areas where lions are actually increasing in numbers, especially in fenced areas. Understanding these population nuances is vital if we are to take the appropriate conservation measures. It is also an important part of gaining a better understanding of the threats that lions pose to people. The occurrence of lions across a landscape and the threats those particular lions face are important in building up an understanding of how the interactions between us and them could lead to human predation.

To return to your imagined life in Tanzania, the fear of predation is high in your mind because you are perched just above the ground trying to protect your crops from rampaging bush pigs. Pigs, you'll remember, attract lions and not only do you live in the country with the highest population of lions anywhere in the world, you live in a region with a relatively high number of attacks; and you are undertaking an activity that puts you at particular risk of being attacked. We know this because, in 2005, well-known lion researcher Craig

Packer and others analysed lion attacks that were reported in Tanzania in the 15 years following 1990.[1]

Lion attacks in Tanzania

In those 15 years, more than 563 Tanzanians were killed by lions and at least 308 were injured. These combined figures were updated in 2007 to exceed 1,000 people attacked.[2] Forty-five per cent of those attacks happened in just six coastal districts and Lindi district accounted for around half. The attacks were characterised in many cases by an unusual feature: lions entered human settlements and areas of agriculture seemingly to seek out humans.[3] In other words, this was active 'deliberate' predation. The number of reported attacks increased quite sharply after 1990 from fewer than 30 per year to around 100 per year from 2002 onwards. 1999 was a particularly bad year, when more than 130 people were attacked. Packer *et al.* attribute this striking increase in attacks to the rise in Tanzania's human population and a linked decrease in the availably of prey for lions outside of protected areas. Where wildlife is protected in national parks, or in the better-protected and managed hunting blocks that spread out across Tanzania's landscape, lions have access to non-human prey. Where lions roam in less well-protected landscapes their normal prey is depleted. In such situations, humans (and that would include you perching above your fields) become acceptable prey.

If we think back to Harry Wolhuter in Chapter One, the first game ranger of Kruger National Park, then his near-fatal attack by lions probably summarises most people's impression of a 'classic' lion attack. He was out on patrol when he was charged by the initial lion. He was on horseback and very much in the 'wilderness'. If we think of predator attacks, it is likely that we think about these wilderness-located, adventuresome encounters. But the data from Tanzania show very clearly that the reality of lion attacks is generally far more mundane. Packer *et al.* were able to categorise attacks into different contexts, and these tell a story of domesticity and village life rather than derring-do and misadventure. Attacks occurred when people were in or by their house, when they were going to the toilet (most commonly outside), or when they were walking, herding or tending crops. In other words, attacks happened when people were living their everyday lives, going about perfectly ordinary activities. Some attacks did occur when people (mostly men) were hunting lions that had attacked people or cattle, but these were the minority. As the paper starkly relates, 'Lions pull people out of bed, attack nursing mothers and catch children playing outside.' To be attacked while outside is awful, but to be attacked by a 150kg (330lb) lion while in your house, perhaps while even asleep, is truly the stuff of nightmares. The construction of rural houses is such that they provide only the merest security. With the majority having roofs thatched with grass or reed stems – and many also having similarly thatched walls – lions can simply force their way inside. If people venture outside to go to the toilet they are especially vulnerable. More than 18 per cent of the 538 victims of attacks whose age was known were under 10 years old.

Time of year plays a part in lion attacks in this area. Nearly 40 per cent of attacks occurred during the

March–May period when people harvest their crops. You will recall that it was April when you were imagining being out in the fields, and you will also recall that you were protecting your crops from bush pigs. The sense of security that a wooden structure beneath you and thatched walls and a roof above might provide is illusionary and your position in the fields makes you especially vulnerable to attack. More than 27 per cent of attacks involve people in precisely your current position, often while they are sleeping.

Complex interactions

The relationship between crop protection against bush pigs, lion presence and subsequent attacks is complex. Packer *et al.* interviewed people who reported that lions were seen entering villages or fields in pursuit of bush pigs. They also reported in some cases tolerating the presence of lions because they helped to control the numbers of bush pig crop raiders, although lions in pursuit of bush pigs may find themselves in villages whose crops attract the bush pigs in the first place. The situation is further complicated by the overall abundance of the species that would normally be hunted by lions. Packer *et al.* concluded that attacks on people were highest in districts with the lowest abundance of species such as kudu, zebra, hartebeest, impala or dik-dik, and the highest abundance of bush pigs. Nearly half of the variance in attacks found between districts (some districts having high numbers of attacks, for example, and others having low numbers) could be explained just by the dearth of prey and the abundance of bush pigs. Interestingly, adding in other factors like human population density, cattle density, agricultural land cover or proximity to a protected area (where lions are more common) did not add any additional explanation for the variance seen in attacks between districts. The conclusion is that lion attacks are being driven by a depletion of prey species and an increase in bush pigs, which are acting as a maintenance diet in areas where settlement and agricultural disturbance have reduced other prey. Bush pigs are attracted to crops at harvest time, so lions follow them. When you add people into the mix, the result can be tragic.

Further analysis of attacks published in 2010 revealed finer-scale patterns in human–lion interactions. This analysis, led by Hadas Kushnir, compared attacks between the Lindi region and the Rufiji district. Rufiji is in the west of Tanzania, is less densely populated than Lindi and contains part of the Selous Game Reserve, a source of wild lions. Overall, Rufiji contains more lions, more prey and slightly fewer people than Lindi in an area a little under 50 per cent larger. The pattern of attacks between the two regions is quite different. Lindi (more people, more densely populated, fewer lions, lower prey) suffered 190 attacks between 1990 and 2007, and Rufiji (more lions) had 101. Interviews were conducted in the two regions, focusing on a pair of neighbouring villages in each region, one of which had high attacks and the other no attacks. Through this approach, the team was able to unpick some of the human activities that made attacks more likely. It confirmed the findings of Packer *et al.* that lower wild prey density and increased abundance of bush pigs were risk factors for lion attacks, but also added the ownership of fewer assets, poorly constructed huts or houses, having to walk further to get resources and more nights sleeping outside. What these additional risk factors point to is the role of poverty in being attacked. Having a poorly constructed dwelling in a location further from resources and owning few assets make you vulnerable. As we will see with other predatory species, it is the poor of the world that overwhelmingly bear the brunt of predation.

The approach used by Kushnir *et al.* allowed them to look at differences between Lindi and Rufiji, and start to disentangle some of the complex human factors that underpin different susceptibility to lion attacks. Clearly, to be attacked by a lion you need two things in place: people in positions of vulnerability and lions; just having lions in an area is not sufficient. What Kushnir *et al.* found was that different human behaviours – some of which link closely to economic development, underpinned by landscape and agricultural differences – were critical in explaining patterns of predation on humans.

In Rufiji, the majority of attacks occurred at night, inside structures in agricultural fields in which people were sleeping – much like the makeshift hut you found yourself in when imagining you were a Tanzanian villager. The Rufiji River separates settlements on the north side from agriculture to the south. People tend to have houses in the villages on the north side and temporary dwellings in their fields on the south side, where they spend most of their time during harvest seasons. Staying in the fields overnight is vital because many of the crop-raiding species – including bush pigs, warthogs, vervet monkeys, baboons and elephants – often come at night. Interviewing the villagers led to suggestions that lions are predominately found to the south of the river and are to some extent unable to cross the river and enter villages. The situation in Lindi is similar in that residents are reliant on small-scale agriculture. However, in Lindi there is no geographical separation between fields and villages, which tend to be closer together than in Rufiji. Lindi residents may have to walk anywhere between 5 minutes to 2.5 hours to travel between the two. For longer trips, overnight stays may be preferable and although Rufiji residents may sleep in fields during some times at harvest, they do so far less than in Lindi. Further differences between the two locations are revealed when residents were asked about water collecting. The river in Rufiji is a convenient local resource, but Lindi residents had to walk, sometimes for an hour or so, to get water from wells. So while in Rufiji attacks often occurred at night in fields, in Lindi people tend to be more susceptible to attacks when walking.

The numbers of people attacked and killed are horrific enough, but behind them lie terrible stories of human suffering. As Amy Dickman succinctly summed up when I interviewed her for a BBC radio documentary on predation of humans,[4] these attacks are 'totally devastating for everyone involved'. Men are taken more often than women and very often those men will be the breadwinners of the family. Children are also killed, causing unspeakable pain for the families involved. While some attacks can plausibly be thought of as defensive, the entering of homes where people are sleeping can only be interpreted as a predatory behaviour. Of the more than 1,000 Tanzanians attacked between 1990 and 2009, more than two-thirds were killed and the victims eaten. That humans are prey in these situations is further underlined by the fact that in the Ruaha district of Tanzania there is evidence of lions actually adjusting their behaviour to take advantage of the novel prey resource offered by humans. Dickman described to me how, when analysing attacks in that area, she had imagined they would be focused on what might be described as 'easy prey': the elderly, the infirm or the very young. In fact, attacks in that area were focused on men aged between 20–40. The vast majority of these men had a common feature that overcame their physicality to make them unusually vulnerable: they were drunk. These men, walking home after an evening of drinking, were blissfully unaware of their surroundings and as vulnerable as drunk people walking anywhere. With a characteristic smell, gait and sound, it is not hard to imagine lions swiftly learning that such people are easy prey.

Another factor in the predation of humans – and in predation overall – is the lunar cycle. Despite the impression created by the countless sequences of lions hunting in daylight on nature documentaries, lions are essentially nocturnal predators and the majority of human predation events occur at night. It gets more nuanced though. Looking at the timing of predation events in relation to the lunar cycle revealed that most attacks on humans occurred in the weeks following the full moon.[5] During this period we get the darkest evening hours, but this is also the time of night when we tend to be most active; a bad combination for us if nocturnal predators are around.

Understanding the patterns that lie beneath human–wildlife conflict is more than just an academic exercise. Knowledge of how, where, why and when attacks happen, as well as who is attacked, allows us to develop and plan mitigation strategies that protect people. Such strategies can also protect predators. Retaliatory and pre-emptive killing based on fear of future attacks is a major driver of predator decline in many cases. I'll return to this topic later on in this chapter when I consider mitigation and the future.

Beyond Tanzania

With the highest number of lions of any country, a growing human population and a relatively well-developed history of wildlife research, it is perhaps not surprising that Tanzania has become a focus of study into lion attacks – but attacks do occur elsewhere. At the end of 2020, news media sites around the world were abuzz with the story of wildlife researcher Gotz Neef. Neef was awoken by a lion while camping in the Okavango Delta in Botswana. The old male lion, described variously as emaciated and starving, had apparently been ejected from his pride and saw Neef as an easy meal. The lion made what turned out to be a fatal mistake when it pounced on the researcher's tent. Neef fought back (as prey often does), punching the lion in the face, while others nearby threw elephant dung and a flash-bang (a loud, firecracker-type device designed to scare away animals) at the lion. The lion retreated back into the bush and was later euthanised (a common euphemism for shot). During the attack, Neef sustained serious injuries including 16 puncture wounds, some broken bones and deep scratches on his head and back.[6] The wildlife researcher survived the attack, but there can be little doubt that had he not fought so hard and had campmates nearby to help, he would have ended up being eaten.

If we look behind the lurid headlines, the story of Neef illustrates a number of important points. First, as with Harry Wolhuter in Kruger, fighting back against a predator individually can only go so far. In the end, both men owed their lives to the support of others. In Wolhuter's case, it was the men who came to rescue him when the first lion that attacked had him stuck up a tree, injured and certain to die without medical intervention. For Neef, it was the quick thinking and bravery of others in the camp that drove the lion back into the bush, and doubtless also helped in the administering of essential first aid. On our own, against a far heavier, stronger and quicker predator we are not physically much of a match. Had Neef been carrying a suitable firearm – and been able to use such a weapon effectively in a highly stressful situation – then he might have been able to subdue the lion without assistance, but it is clear that, even then, having others around would be a strong advantage. Our sociability and our communities protect us to some extent against predation.

The second point we can take from Neef's encounter is an idea we will return to frequently in other chapters: the 'incapacitated man-eater'. The lion that attacked

Neef was described as old, emaciated and starving, and certainly such an animal would have problems taking an antelope, warthog or bush pig. I saw a highly emaciated lion in Pilanesberg National Park in South Africa in 2018 that was most likely in a similar state to the one that attacked Neef. The animal walked slowly, pelvis and spine clearly visible through the skin. No longer part of a pride, and with no realistic chance of catching anything larger than a dung beetle, this individual will likely have either starved to death or been taken by hyenas, or other lions. It is easy to imagine such an animal desperately wandering into a human camp in search of an easy meal. Indian-born British hunter and naturalist Jim Corbett – who we will meet again in Chapter Three – suggested that most tigers and leopards that resort to hunting people were injured or incapacitated in some way. While that may have been true of the cats Corbett hunted down – and might be true in some cases now – it does not seem to be the case in modern-day India or with the lions attacking villagers in Tanzania.

The third point is that predatory lion attacks happen outside of the relatively well-documented areas of Tanzania. Neef was attacked in Botswana and, albeit more than a century ago, Wolhuter was attacked in South Africa. In Kenya in 2017, 18-year-old Weldon Kirui was attacked and mostly eaten in 2017 in Nairobi National Park,[7] a protected area only 7km (4 miles) south of Kenya's capital city. Nairobi's high-rise skyline is actually visible from much of the park, although less so at 2 a.m. when the attack happened. The victim was a Maasai who was grazing his cattle in the National Park at night, which is illegal but a drought had forced the herders to take advantage of whatever pasture they could find.

It is instructive to look at the way the attack on Kirui was covered outside Kenya. The only mention I could find online in a UK press outlet was in the *Evening Standard*.[8] The details of the attack are related, but Kirui, despite being named in the Kenya report, is unnamed. He is simply identified as a Maasai, stripped of any personal identity. This fatal attack, barely mentioned outside Kenya, contrasts sharply with Neef's non-fatal attack, which was covered globally. Neef, a white researcher on an expedition funded by *National Geographic*, was always named. Neef was never just 'a wildlife researcher' or 'a man'. Meanwhile, if you search for 'Nairobi lions' in the UK press there are plenty of mentions of a lion that escaped the National Park and was shot, replete with plenty of angry comments from readers incensed that a lion was killed. The death of Cecil the lion was a global phenomenon, yet the deaths of countless rural poor barely warrant a mention. It seems that not all Black Lives Matter.

This skewed priority, where individual animals are accorded distinctions, rights and protections above those of the rural people living 'elsewhere' – or where the lives of white people from developed countries are given higher priority than others – is a major issue in conservation. It is far from a recent phenomenon. Kenya was home to perhaps the most famous 'man-eating lions' of all: the Tsavo man-eaters. This coalition of two male lions terrorised the workers constructing the railway through Tsavo between March and December in 1898. The death toll inflicted by this pair has been variously disputed, with estimates of between 28 and 135 victims. Some indication of why it has proved difficult to calculate the number of people killed can be derived from the fact that a well-known account in 1907 gave the death toll as 28 labourers and 'scores of unfortunate African natives'. The name of the man who killed the lions, Lieutenant Colonel Patterson, is well known and accompanies every account of the incident (including this one) and yet the names of the victims are largely forgotten. Even in modern times it is difficult, and often impossible, to get reliable data on the number of people killed by predators when such attacks happen to the rural poor in relatively underdeveloped regions.

As locations for lion attacks in the modern world, we can add South Africa to Tanzania, Kenya and Botswana. There are a number of reports of (again, unnamed) poachers being killed and eaten by lions in Kruger National Park, sites in Limpopo province and other locations. People are also attacked and killed by lions in South Africa, but not eaten. In these cases, it is not always clear whether such attacks are truly predatory. Kobus Marais, an anti-poaching ranger and dog handler I spoke to in 2016 while fact-finding for a BBC Radio 4 documentary on rhino poaching, was killed by a lion in Pilanesberg National Park in 2021. The lion that attacked him was officially described as being very thin, emaciated and in poor condition (similar to the lion I saw in the same location three years previously). The lion was shot, but not before Marais had suffered fatal wounds from the lion, which was described as 'lying in wait'.[9] All the evidence in that case pointed to Marais being a victim of a predatory attack, even though he wasn't eaten.

In an international seminar on the conservation and management of large carnivores in Africa, held in 2006, the session covering lions, conflict and conservation stated that 'even in the twenty-first century man-eating is a serious problem in Ethiopia, Tanzania and Mozambique',[10] adding two more countries to a growing list. Mozambique is home to a growing and viable population of lions, but also suffers from attacks. For example, according to a report prepared in 2007, since 1974 at least 34 people had been killed by lions in the Niassa National Reserve.[11] During just the 27 months between July 2006 and September 2008, 24 people were reported to have been killed by lions in Mozambique, with a concentration of attacks in the area bordering Kruger National Park in South Africa.[12] Ethiopia saw a particularly striking and horrific example of lion attacks in 2005 when, according to reports, a pride of lions killed 20 villagers and injured 10 more in a week.[13] More recently, a lion pride killed four people in 2020 in the Gambella region and reports of those attacks include reference to similar events in previous years.[14]

We can also add Uganda to the list of countries with well-documented lion attacks. Data maintained by the Ugandan government showed that there were 275 attacks by lions on humans there between 1923 and 1994, and 75 per cent of attacks were fatal.[15] Elsewhere, in 2017, a 10-year-old girl going to the toilet was killed by a lion in Zimbabwe,[16] while three boys were lucky to survive an attack there in 2020.[17] In Zambia, lions killed and ate

Lantone Phiri as he was walking in the remote town of Nyimba in 2011,[18] and 21-year-old Anthony Chibulu was killed and partly eaten in the Mambwe district. Zambia was also home to one of the largest human-attacking lions ever recorded, the Mfuwe man-eater,[19] which killed and ate six people in the Luangwa River valley before being shot in 1991.[20] There are many such stories, both reported and unreported.[21]

Most people living in lion country will not end up being attacked and may only rarely, if ever, see a lion. Nonetheless, the presence of lions is a prerequisite for being attacked. Outside of the countries where lions are relatively more numerous, attacks are much more unusual. The rarity of lions in these areas – and the fact that lions may be restricted to known and protected areas (and so less likely to be free-ranging) – probably plays a role in reducing attacks. In West and Central Africa, for example, where lions are found in low densities in countries like Benin, Niger, Cameroon and Guinea, deaths of humans are rare.[22] However, the infrequency of lion attacks on people does not mean that conflict with lions, especially through the depredation of cattle, is not a problem in these areas – as we will see shortly.

Beyond Africa
The only place outside Africa with free-ranging wild lions is Gir National Park, a forested area in Gujarat, India. According to official figures (and keep that 'according to' in mind as we move ahead), this population has been increasing at an accelerating rate in recent years, and numbered more than 674 in 2020 (a rise of 29 per cent from 2015).[23] Prime Minister

Narendra Modi tweeted this good news and announced that the population had also had a geographical range increase of 36 per cent. This conservation success story was widely picked up by the world's media, as indeed it should have been. However, a bit of digging reveals that the situation is a little more complex than Modi's tweet suggested. It was by no means all good news for the lions of India during 2020. Across the year, despite the increases trumpeted by Modi, lions started to die and in numbers that were, for some observers at least, concerning. Ten lions died in January, a further 12 in February, followed by 10 more in March. By the time May ended, a further 50 were reported to have died.[24]

Something that can be forgotten when media stories document 'mass die-offs' and other seemingly shocking mortality events is that animals die. Nothing lives forever. When you have a reasonably sized population of animals that live, as lions do, for perhaps 10–14 years, then you would expect some turnover. Cubs will be born and a proportion of these will not make it to adulthood. Meanwhile, older lions will die. The issue, with any population, is when the death rate observed exceeds what might be reasonably expected to occur from the natural background mortality rate. When this point is breached it becomes reasonable to assume that something untoward may be going on, and that some intervention may be required to stop or slow down the additional population loss. This is especially the case for species of conservation concern and when we suspect (as is often the case) that the increased mortality may be due to something we are doing. In fact, lions around Gir have died from unnatural causes before. In the areas around the protected National Park, where lions can roam in a wider human-dominated landscape, at least 30 lions died in a 10-year period as a consequence of falling down open wells. When deaths occur as a consequence of some human-centred activity or feature then it may be relatively straightforward to prevent or reduce further deaths. In this case, the simple measure of building a parapet wall around wells was sufficient to result in no lions, or any other species, dying in them.[25] Of course, with tens of thousands of wells already in the landscape and others being dug, a wholesale conversion is by no means straightforward, but it is at least possible. Other sources of mass mortality can be far harder to ameliorate, even if the source of mortality can be determined.

The cause of the lion deaths occurring in 2020 has so far proved hard to pin down. Indeed, there seems to be little agreement as to whether the deaths even require an explanation. In March 2021, the Gujarat government announced in the state assembly that 313 lions had died in the Gir National Park in the two previous years,[26] but only two adults and eight cubs were considered to have died from 'unnatural causes'. In 2019 the state assembly was told that as many as 222 lions died after 2017, but that only 23 of these had died from unnatural causes (which included falling into wells).[27] At this point, politics kicks in. The opposition party claimed in 2020 that rotting cattle meat, transported illegally into the National Park from surrounding villages, was to blame for deaths that were being counted as natural. Meanwhile, canine distemper virus (CDV), which is known to have caused lion deaths in Gir in 2018,[28] has been suggested as a cause of excess deaths by some. A reasonable explanation – CDV and related viruses have caused mortality in African lions – but its presence in India was denied by the chief conservator of forests of the Junagadh Wildlife Circle, D. T. Vasavada, who stated in 2020: 'There is no CDV here in Gir ... This issue of CDV is a media-versus-Gujarat government thing; it has no truth to it.'[29] In other words, CDV was fake news, with a whiff of political game-playing. It is a measure of our relationship with lions, and their cultural and even political importance, that they are affected by fake news and political posturing.

The situation in India is further complicated by the fact that counting lions is far more difficult than it first appears. Bear in mind that in India lions only live in one area, centred on Gir National Park. That is an area of 1,412km^2 (545 square miles), supporting 674 lions, according to the 2020 figures. This is an area just 7 per cent the size of Kruger National Park in South Africa (19,485km^2/3,662 square miles), which is estimated to have around 1,600 lions. Although very different types of habitat, having nearly half the individuals in an area almost 14 times smaller, seems as though it should make counting lions in Gir relatively straightforward compared with other areas of lion range. Once you add the fact that lions are sizeable creatures, often quite vocal, live in groups, and leave kills and footprints around the landscape, you start to build up a picture of an animal that should be relatively easy to census. As is often the case when armchair assumptions get applied to ecological scenarios, the reality is rather different. Even with healthy populations in smaller areas, lions are still relatively thin on the ground. If we assume 674 individuals evenly spread through the 1,412km^2 (545 square miles) of Gir National Park, then that is only one lion for every 2km^2 (0.7 square miles). Even at relatively higher densities, the lions of India have a lot of land in which to hide. That is, of course, before you include vegetation cover and topography. Lions are also surprisingly well camouflaged in natural vegetation. That nondescript light brown, in conjunction with the ability to be very still while lying down, can make lions all but invisible. I once drove past a vehicle in Botswana that had stopped to view a lion resting just 15m from the road. The occupants had got lucky and glimpsed an ear moving slightly as they drove along. Coming in the opposite direction, we certainly would not have spotted it. They left us to the sighting and we watched this lone lion for a few minutes before realising that there were other lions lying up in the grass. In fact, at least four other lions were eventually visible and I am quite sure there may been others sleeping in the hot afternoon sun. The reality is that animals can be very hard to count even with the best methods available – and in Gir National Park it has been suggested that the methods being used to count lions are far from the best available.

In 2020, Yadvendradev Jhala, senior scientist at the Wildlife Institute of India, was damning in his appraisal of the lion-survey methodology being used in Gir National Park. Dismissing them as 100-year-old methods that would not 'stand the scrutiny of contemporary science', Jhala makes a strong point. The lion survey was conducted using the 'block count' method, which is logically very straightforward. You count lions that you find in given survey areas (the 'blocks'), then multiply to estimate the lions that you should find across the whole area. In this case waterholes within counting blocks across the park were staked out for two days and lions were counted by 1,400 people. There are all kinds of problems with this approach, but often in ecology you have to work with what you have even if it is far from ideal. The accuracy of the final estimate produced by this method is greatly affected by the behaviour of the animals; for example, if they roam over large areas the same animals can be counted in multiple blocks and lead to an overestimate. Alternatively, a failure to detect animals at the survey point, if they happen to avoid waterholes that day, for instance, can result in underestimates. These problems are well known, and modern methods – including using camera traps, the identification of individual animals by unique marks or from DNA analysis, sophisticated statistical approaches and greatly increased survey efforts – generally provide far more accurate population estimates. However, some experts have argued that camera trapping is not effective for lions because they are harder to identify individually; unlike leopards or tigers they don't have so many obvious, individual-specific markings. Estimates from some quarters suggest that there are probably 700 more lions living outside the protected survey area (although it is unclear how these estimates are reached).

It is all a bit of a mess. I contacted Jhala in an attempt to get some clarity but, alas, his reply only made things murkier. As he told me in 2021, 'There could be anywhere between 350 to 1,000 lions. Without a proper assessment, numbers are political populations.' Meanwhile, the issue of whether there are in fact 'excess deaths' and unnatural mortality – and if so whether these deaths are caused by canine distemper virus or some other factor – remains unresolved and hotly contested. Again, I was hoping for clarity from Jhala, but his frustration with the situation is clear in his reply: 'There has been no transparency regarding CDV and lion deaths. My research permits of 25 years were cancelled, since what I have been saying and writing based on scientific data is not palatable to the forest officials and the government (I work for the government, though!). I do not have scientifically valid information on CDV deaths, but lions are dying in the Gir landscape.'

What is a little clearer is the threat posed to people by lions. Between 2007 and 2017 in the wider Gir landscape, 190 attacks were recorded resulting in 12 fatalities (4 per cent of attacks or 1.3 deaths per year). For comparison, in the same landscape, leopards were responsible for 383 attacks in 2011–2016, resulting in 41 fatalities or around 7 per year.[30] The low percentage of fatalities from attacks is interesting and the conclusion is that lion attacks are 'mostly accidental', arising not from deliberate stalking of humans as prey (the usual pattern in Tanzania), but from self-defence when lions are 'spooked' by surprise encounters. Revealingly, an attitudes survey of residents in the area showed that communities benefiting economically from lions (by lion predation on agricultural pests like bush pigs and from tourism) were far more tolerant of their presence than pastoralists who experienced some livestock losses from lions. This is an important point to bear in mind and one that we will return to frequently. People can be surprisingly tolerant of dangerous predators if there is some benefit to them being around. At this point – and I am assuming a few things about you as a reader – it would be helpful to park your privileged notion of wildlife having aesthetic or intrinsic value. Unless you live in fear of being stalked by a lion, or having your livelihood destroyed by one, you don't really have a solid position to comment on costs and benefits.

Living with conflict

There is a great deal of uncertainty when it comes to gauging the precise impact that lions have on humans through direct, predatory attacks. Record keeping varies, data may not be collected well or at all, deaths may not be recorded, bodies may not be recovered and what information we have is often distributed among different sources. As such, it is currently impossible to say exactly how many people are killed by lions every year across their entire range. What we can say, with certainty, is that lions do present a real risk to human beings through predatory attacks in many areas where the two species coexist. Lions do hunt and kill people, and the rural poor in countries like Tanzania bear the brunt of these attacks.

Simple coexistence between lions and humans is not sufficient to elevate the threat from 'notional' to 'real'. Lions in protected areas, where human–lion interactions may be more controlled, represent far less of a threat than lions that roam freely in areas where people live, grow crops, walk or work. However, it is also fair to say that lions in these areas have far more to fear from

people than people do from them. This is not to trivialise
the fact that hundreds of people a year are injured or
killed, and eaten, by lions. But from a conservation
perspective we cannot lose sight of the fact that a great
many lions are killed in retaliatory and pre-emptive
attacks.

While the killing and eating of people clearly
represents the extreme end of human–wildlife conflict,
it is still relatively uncommon compared with other
forms of conflict. With lions, this conflict is most often
manifested through the killing of livestock. Predation
of livestock is a common concern for many who live in
areas where predators are present. Plans to reintroduce
predators like wolves to landscapes, including part of
Scotland, are often met by objections based on the fact
that predators kill livestock (see Chapter Nine). It is
common to hear these concerns downplayed by
advocates of 'rewilding' and certainly it is by no means
a given that predators in areas with livestock will
become 'livestock killers'. However, predators most
certainly do kill livestock and for poor people living
on marginal land such losses can become life-
threatening. It is no surprise that people faced with
potential losses will not spend much time evaluating
relative risk when a simple solution is to hand.
Consequently, lions and many other predators have
long faced persecution from people scared for their
lives and the lives of their animals, which for many
communities around the world are deeply intertwined.
A common theme that emerges, whether it is goats in
Tanzania, cattle in India or sheep in the Highlands of
Scotland is 'compensation'. Sure, you might lose some

animals, but 'we' can just pay the bill and move on. I would suggest, politely, that it is a very privileged position indeed to pontificate about compensation for livestock losses when you neither suffer the loss nor write the cheque.

Reducing conflict

Reducing conflict is not straightforward, but before considering the options available we need to examine the nature of human–lion conflicts more generally. This involves tackling some difficult topics that lurk beneath the surface of many conservation issues. My focus in this book is on human predation and it is inevitable perhaps that much of the negative lion media coverage highlights these attacks. As we have already seen, though, media coverage of attacks is far from uniform. The non-fatal mauling of a white man in Botswana can become viral news, while the killing and eating of villagers in Ethiopia escapes any international attention. We should not then be surprised that when most lion-related deaths do warrant a mention, their coverage is usually derisory at best. By not naming victims, such coverage perhaps also reveals deep racial prejudices and the persistence of a pervasive, and largely unchallenged, colonial attitude when it comes to predators and the people who live alongside them.

In fact, the developed world media tends to be far more focused on the predators themselves than the human cost they impose. When Cecil the lion was shot and killed by Walter Palmer in 2015, the story took a few weeks to really take hold – but once it did it become a popular news story for days and then weeks. The media outrage – and subsequent public outrage – grew rapidly, but I saw no reports of anyone being killed and eaten by lions in the UK media during the 'Cecil summer'. There most definitely will have been such incidents, but the death of poor rural people in distant countries seems to resonate far less with Western observers than the death of a lion.

David Macdonald, the then head of the unit (WildCRU) at the University of Oxford that had collared Cecil, said that 'in terms of attracting global attention, it [Cecil] was the largest story in the history of wildlife conservation'.[31] Let's just let that sink in. We are seeing habitat destruction across the world; poisoning incidents that involve hundreds of vultures at a time; a rhino-poaching crisis in South Africa that is nowhere near under control; an international illegal trade in pangolins that many believe is an existential crisis for several species; the bleaching of coral reefs; over-exploitation of marine resources; the list goes on. Against this backdrop, a leading conservation biologist identifies, rightly in my opinion, the death of a single, old, lion as 'the largest story in the history of wildlife conservation'. As I mentioned at the start of the book, I visited South Africa to record a BBC radio documentary about trophy hunting around two months after the Cecil story broke, and Cecil and Palmer were both still big news, with stories, updates and speculation appearing frequently in both traditional and social media. I am typing this on a morning more than six years after the event and a quick search of Twitter reveals 11 tweets in the past 24 hours directly referencing Cecil the lion, all of which have other Twitter users engaging with them. A single lion, that died six years ago...

Some balance on the Cecil story was provided by the Zimbabwean media, although the voices heard there were seldom amplified internationally (see the pattern?). Alex Magaisa, writing in the Zimbabwe paper the *Herald* on 30 July reported that he had never heard of Cecil and that neither had his friends and family. Magaisa went to say that 'the manner in which the story has been presented by international media seems somewhat far removed from the lived realities of most of the local people'.[32] As Keith Somerville pointed out in his definitive paper on the coverage of Cecil,[33] the Zimbabwean columnist Farai Sevenzo wrote on BBC's African news website (interestingly on the same day that Magaisa was commenting in the *Herald*) that 'Zimbabweans feel somewhat bemused by the attention the world is giving to the killing of a lion'. What Sevenzo goes on to say is especially relevant when it comes to addressing wider issues of human–wildlife conflict. He points out that Zimbabwe was confused by the surge in international media interest in Zimbabwe, especially as it 'did not come from the high unemployment figures, the food shortages, the state persecution of vendors, the lack of medicines, the lack of cash – but from a lion named 'Cecil' by conservationists'.[34] In other words, Western observers seemed to value the life of a single animal over the lives of Zimbabwean people.

Cecil the lion was perhaps an aberration, a coincidence of timing and events. The viral nature of the story can arguably be traced to what the *Daily Telegraph* described as an 'impassioned rant' by US TV talk-show host Jimmy

Kimmel.[35] During a section of the show focusing on the
story, Kimmel asked an absent Palmer, 'The big question
is, why are you shooting a lion in the first place? I'm
honestly curious to know why a human being would be
compelled to do that. How is that fun? Is it that difficult
for you to get an erection that you need to kill things?'
Kimmel also appealed to people to donate to WildCRU.
The conservation unit's website crashed and it received
more than a million dollars during the weeks following
Kimmel's appeal. Whether Kimmel would have
encouraged such donations knowing that members of
WildCRU – as conservation scientists and leading experts
in lion conservation – have publicly, and patiently, sought
to explain the evidence as to why trophy hunting can in
fact be good for conservation is a question I will leave
hanging. With a large audience and a wide sphere of
influence, Kimmel's outburst, during which he appeared
to shed tears, gave the story a boost that the internet and
social media subsequently magnified. Further stories
involving named animals and hunters began to emerge,
including Xanda the lion ('son of Cecil')[36] and Voortrekker
the elephant,[37] but none of the sequels have had the
traction and longevity of the original.

When media reports do carry stories of predator
attacks on people, those reports can be equally unhelpful
for conservation. By ignoring those who are most likely
to be the victims of attacks, the media blur the public
focus on real conservation issues; but by sensationalising
attacks people's natural – and sensible – wariness of
predators is enhanced to a point when these animals
become feared and vilified. In a study of 1,774 media
reports of large predator attacks most stories (59 per

cent) were published in North America, followed by
Europe (20 per cent), Asia (15 per cent) and only 2.6 per
cent in Africa despite a large number of attacks by lions
and crocodiles (more on crocodiles in Chapter Four).[38]
Lions barely featured and it was brown bears (which
we'll cover in Chapter Eight) that led the way, accounting
for 16 per cent of stories. The number of people killed in
a week in Tanzania alone is probably double the number
of people killed each year by brown bears in North
America, but you wouldn't know that if you focused on
media reports. The language being used to describe
attacks – and the predators themselves – was often
negative and included words like 'horror', 'nightmare',
'terrifying' and 'gruesome'. Graphical elements including
lurid descriptions and photographs are often present,
and may be enhanced on social media where such
elements can also encourage wider sharing. The UK
press rarely shows graphic images, but other countries
often have fewer reservations. It still comes as a shock
seeing front pages abroad featuring human remains,
gory injuries, mutilated bodies and so on. As I will talk
about in the next chapter, I frequently receive horrific
images from India of people who have fallen victim to
tigers and leopards, and in many cases these images
have appeared in the local press or are shared on social
media. Such images and loaded language might, as the
authors of the study point out, 'lead to an unjustified
and amplified fear in the public with consequent lower
tolerance toward predators and decrease in the support
for conservation plans'.

A fear of predators is entirely rational and reflects an
evolutionary legacy from our long line of ancestors who

faced a real and regular threat of becoming prey. I
interviewed a number of researchers who work with
predators for a BBC radio series and a common theme
that emerged was the 'hairs on the back of the neck
standing up' reaction that we have to them.[39] For many
of us living modern lifestyles the actual threat posed by
predators is essentially zero, but the threat posed by
predators to many in the developing world is very real.
Interestingly, the perception of the risk of attack is also
very real, at least for those living in Tanzania.

A study of the perception of risk of predatory lion
attacks for people living in Tanzania was published in
2019 by two researchers we have met before: Hadas
Kushnir and Craig Packer. As in previous studies we
have explored, it was the districts of Rufiji and Lindi
that came under the microscope, this time through
questionnaire-based interviews of members of
randomly selected households. The study included
questions that aimed to unpick the effects of
demographic factors (age, gender), personal experience
of lions, and family history of attacks on perception of
risk. Two questions specifically asked people about their
perception of risk ('How likely do you think you are to
be attacked by a lion?') and concern about that risk ('Are
you worried about being attacked by a lion?').[40]

As we know, more than 1,000 people were attacked by
lions in Tanzania between 1990 and 2007. Most attacks
were unprovoked and clearly predatory in motivation.
The numbers are high and we must not lose sight of the
human cost, but equally the overall risk to any given
individual is low. Kushnir and Packer conclude that
with 'an average of 15.5 attacks per year in Rufiji and

Lindi, a combined population of ~450,000 people in the two districts and an average lifespan in Tanzania of 55.9 years, a realistic estimate of an individual's lifetime chances of being attacked is well below 1 per cent'. This is true when we consider the population overall, but not everyone in these districts is equally at risk from attack. We know that certain activities put some people at more risk than others, so the risk posed by lions to certain groups within a population is higher than the calculated population-wide risk; likewise some groups have lower risk. We can also be swayed by seemingly low numbers. Less than 1 per cent might be a totally acceptable risk for, say, suffering a sprain playing tennis, but a wholly unacceptable risk for something that has potentially fatal consequences. Our appetite for risk depends on the nature of the consequences of that risk. In June 2021, I completed the QCovid University of Oxford risk calculator,[41] and based on a number of factors relating to my age, gender and health I had a considerably less than 1 per cent chance of even being admitted to hospital with Covid-19 and a far lower risk of dying. Despite this low risk, I was very happy to receive my vaccinations and, I won't lie, that happiness didn't come from the perspective of the 'greater good' and the lower risk of transmission that comes from being vaccinated. Even though I know the numbers, my appetite for disease risk is relatively low, which is also the reason I get a flu jab every autumn. I have a family, dependents, a life that I don't want to risk if I don't have to. Sure, my actual risk is low, but perhaps my perception or inward sense of risk – and the amount I may be concerned by it – are higher. Rural Tanzanians, facing existential threats, have a similarly skewed but understandable sense of risk.

Across the survey, 53 per cent of people thought they were 'very likely' to be attacked by a lion, and 69 per cent worried about being attacked. There were no significant differences between people living in villages who had suffered lion attacks and those who hadn't, or between people who had suffered a family member being attacked and those who hadn't. People with more education were more likely to be worried about attacks and thought they were more likely to be attacked, but there were no differences in perception or concern between genders. There were also no differences based on sightings of lions or signs of lions, with one exception: people who saw signs of lions (tracks and scat) in their village were more likely to be worried or very worried about being attacked than those who hadn't seen signs in their village. And who can blame them?

As might be expected, the discussion section of this study focused on the high perceived risk and level of worry compared with the actual risk. Academic discussions about how psychological factors influence risk perception, the level of loss, false beliefs, mismatches with other higher risks, the role of dread and fear, and so on, are interesting – and potentially important in understanding and mitigating human–wildlife conflicts. Understanding where risk is over- and underestimated, for example, can direct behavioural changes and education towards reducing risk, to the benefit of people and lions alike. What such discussions ignore, though, is a very basic and far more human fact. A good deal more than half of the people living in these districts live in fear of being attacked and eaten by lions.

Reducing the fear and worry associated with living alongside predators is not an easy task. Ultimately it may be impossible to remove concerns without removing all predators, no matter how small the actual risk may be. It is at this stage that we need to accept that grown-up decisions need to be taken if we want the world to be a certain way. We cannot simply look on from afar and decide that lions are more important than people, and that losing a 'few people here and there' is an acceptable price to pay. This sentiment, rife on social media among those who profess to love animals, is inhumane, ignorant and hugely damaging for conservation. The simple fact is that conservation goals of maintaining predators in the wild may well be in conflict with local goals that prioritise human needs. It is finding the delicate balance of these often-opposing forces that is the essence of modern, informed and successful conservation, but it is far from easy. To be successful in maintaining predators around human populations, where the risk of being attacked and eaten are real, we cannot ignore people, but if we can keep people safe then we also stand a good chance of keeping predators safe.

One way to reduce the risk of predation is simply to fence predators into areas in which humans are excluded or allowed to venture only in very controlled ways. This 'fenced fortress' model of conservation may also be successful in reducing poaching and producing workable economic wildlife-based activities like photo-tourism. It can certainly work to reassure communities. When I interviewed people around Pilanesberg National

Park in South Africa about their fears of living with
lions just over the predator-proof fence, I got very little
sense of concern from them precisely because the park
is surrounded by a well-maintained secure fence.
Leopards elicited more of a reaction – and their
reputation for being versatile, wily and stealthy is
certainly well deserved, as we will see in Chapter Seven.
The main concern, though, wasn't a predator but a
herbivore; escaping elephants were far more worrying
to local people than animals with sharp teeth and claws.
It would be interesting to see whether people's fear of
lions increased following the death of Kobus Marais,
but my guess would be not. The incident, horrendous
though it was, occurred within the fence and to someone
working in a job that is inherently risky.

Speaking to Craig Packer in 2015 it was very clear
that he felt fences were important for the future of lions,
keeping them safe from us and us from them. I well
remember standing by a large fence in the Dinokeng
Game Reserve in South Africa interviewing him about
lions with several large males sitting just a few metres
away on the other side of the fence. The lions were part
of Kevin 'the lion whisperer' Richardson's sanctuary,
associated with Dinokeng, and visitors staying at the
sanctuary could have close encounters with his lions.
Richardson and the Dinokeng sanctuary hit the
international news in 2018 for all the wrong reasons.
While Richardson was walking with a colleague and
three lions in the reserve, one of the lions charged after
an impala, ran a couple of kilometres and encountered a
22-year-old woman who was on-site with a friend
interviewing the manager of the Dinokeng Game

Reserve. The woman, who was unnamed in the resulting international media, was described as being 'mauled to death', but given the circumstances it feels safe to describe it as a predatory attack.[42]

Three years before that incident, and no more than a few kilometres from it, I asked Packer what we needed to keep lions safe and his reply was simple: 'Fences, fences, fences', adding later, '... and money'. His answer neatly reflected the title of a paper, of which he was the lead author, published two years before our interview: 'Conserving large carnivores: dollar and fence'. Lion populations in fenced areas were concluded to be more stable and cheaper to conserve than unfenced populations. A key argument made by Packer and others is that fenced populations can reach numbers close to their carrying capacity, whereas unfenced populations rarely do due to human factors. Overall, they concluded that half of unfenced populations are ultimately doomed over the coming few decades.[43] Others, including Amy Dickman, have argued that fenced populations can often exceed their carrying capacity, causing costly issues – something that is widely acknowledged when I have spoken to managers of fenced populations – and are relatively small. Due to these factors unfenced populations allow many more lions to be conserved per dollar; in other words, unfenced populations represent a more efficient use of funds. Fences also impose considerable economic cost initially (predator-proof fences at landscape scales are not cheap) and fragment the landscape, imposing an ecological cost.[44] Fences that keep lions in also prevent the natural movement of other species that are unable to travel under, through or over the fence. This can be a major problem for herbivores especially, which can end up overgrazing areas they are unable to move away from. Over time, populations may also become inbred, reducing genetic diversity, and potentially reducing overall population health and productivity. Aside from the larger and more obvious species, fences can prevent smaller animals – like honey badgers, genets and civets – from moving around the landscape, potentially producing pockets of lower than normal biodiversity. Some of the issues fences cause can be solved with good management. A man with more than 60 years of wildlife and habitat management summed it up to me very succinctly, 'If we put a fence around, we have to manage it,' and management is costly.

There are sometimes further, human, costs to fences and to exclusionary 'fortress' conservation methods. The exclusion of people to create protected areas, whether they are fenced or not, can lead to human rights abuses at all levels, including exclusion from areas of cultural significance and denial of livelihood. Human rights abuses may escalate with this approach, as has been well documented in the Republic of the Congo (and elsewhere), where armed 'ecoguards' – funded in part by the conservation group WWF – beat up, intimated and, according to some reports, imprisoned and tortured Baka tribespeople living close to a proposed national park.[45] There was no actual fence involved, but the exclusionary philosophy is the same. In summary, fences can solve some problems but cause others, and they are a source of sometimes heated controversy and disagreement across conservation.

What we can say for sure, though, is that putting a fence around lions does work in some places, but it may not be an effective or desirable solution everywhere. If ever there was an example of where one size really doesn't fit all, it is conservation.

If physically separating the living spaces of lions and people is not always an option then it may be possible to make people safe from lions in their homes and villages. In principle, 'lion-proofing' dwellings should be reasonably straightforward. If a lion was outside my house right now, for example, I would feel quite safe inside, with doors and windows firmly shut. There is no chance whatsoever of a lion coming in through the walls, crawling up and in under the roof, or simply walking through the entrance. However, my home has a sturdiness and permanence that is not shared, or even possible, for many that live in areas where there are lions. Many rural villagers in Tanzania and elsewhere are living in structures that are not, and cannot be made, lion proof as a consequence of the building materials being used. Traditional mud houses offer some security perhaps, with walls more solid than those in dwellings made from grasses and reeds, but unless entrances can be blocked effectively – and a roof made solid and well connected – it would be hard to call it lion proof.

The main source of human–predator conflict is livestock depredation, and lion killings are often motivated by retaliation for past livestock predation or as pre-emptive killings to prevent future losses. Keeping livestock safe from lions has multiple benefits: protecting livestock from horrendous stress, injury and death; protecting people's livelihood and security; potentially preventing lions from habituating to easily available food in proximity to people; and reducing human–lion conflict and lion killings. One approach that has proved simple and effective is the lion-proof *boma*.

The term '*boma*' (also *kraal*) is widely used across eastern and southern Africa. Most often it refers to a livestock enclosure, although the word can also be used for other forms of enclosures including those where people gather. A *boma* can be as simple as some thorny acacia branches piled up to roughly enclose an area in which livestock can be corralled for the night. Alternatively, it can be used in a grander sense, referring to a stockade or a fortified enclosure. A simple *boma* uses materials found in the bush to create a structure for keeping relatively docile herbivores contained. It is cheap, easy and works well at containing livestock, but it does not keep predators out. In Kenya, cattle tend to be herded into *bomas* at night and when lions approach cattle panic, causing a stampede that results in cattle bursting out of the *boma* and into the bush. Lions and hyenas (see Chapter Six) may take some, and there is considerable cost involved in rounding up the rest. Some lions may even learn to jump over *boma* walls. Collaring studies showed that lions collared on livestock kills were nearly four times more likely to be shot in response to livestock damage than those collared on wildlife kills, strongly suggesting that lions start to specialise in livestock. Lion-proof *bomas* developed around the Amboseli National Park in Kenya by the charity Born Free consist of 2m (6ft 7in) high eucalyptus posts, triple-twisted chain-link fencing and flattened oil drums, and have proved effective at keeping livestock (and in turn, lions) safe,[46] reducing human–lion conflict to the benefit of both species. However, while a simple *boma* is cheap or free, lion-proof livestock *bomas* involve more effort and additional resources.

Lion-proofing *bomas* by using materials not found readily in the bush is a great approach in some areas, but it can't be used everywhere simply because resources are limited. However, thick thornbush *bomas*, or *bomas* that are made from stone (especially if there is a fence on top) can be effective. Another method is to locate livestock *bomas* nearer to humans, where studies have shown lions are generally reluctant to approach, although weaker lions or lions that have become attuned to hunting livestock could potentially develop into more of a problem if livestock and people are located close to each other. Dogs have also proved to be effective at deterring lions, not by chasing off the predators, but by warning herders who can then do the chasing. The fact that lions can be moved away from sites by people giving chase leads to the possibility of 'hazing' predators.

Hazing, or aversive conditioning to give it a more scientific name, is the idea of building up a negative association between a behaviour (approaching livestock for example) and an unpleasant stimulus such as bright lights, a loud noise or being chased. If the behaviour is 'negatively reinforced' enough, then the animal should avoid carrying out the behaviour. To put it another way, you scare the crap out of the animal when it does something you don't want it to do and it learns not to do it. In principle[47] this technique can work well, but in the real-world conditions of the field it can be difficult to pull off. I have seen this first hand with attempts to 'teach' rhino to avoid hanging around near fences where they could be shot from a road by poachers after their horns. Initially, beeping vehicle horns and making a noise is enough to move a rhino on, but fairly swiftly they get used to the disturbance. It isn't long before they park up where they like and ignore pretty much anything. The problem, then, with this approach is what is termed 'habituation', when animals get used to an unpleasant stimulus and it no longer has a strong effect on them. This is what has been found with lions. Aversive conditioning can work – and for some lions it does – but there needs to be consistent aversive conditioning and ideally it needs to occur early on in the development of the behaviour you are trying to condition them away from. When chases were isolated events or performed inconsistently lions did not learn, and in practice it is not always going to be possible to chase lions away consistently enough for them to learn. Nonetheless, this 'hazing' approach may be promising in some situations and is another tool in the human–wildlife conflict-reduction toolbox.

A fatal solution?

At some point it may not be possible to deter a lion that has focused on killing livestock or hunting humans. It may be necessary to kill the lion to prevent further harm. That killing may be undertaken in a variety of ways, some legal and with an eye on welfare, and some less so. Communities may take the decision to hunt down the lion, and in many places this would likely include the use of dogs, spears and whatever else was to hand. When humans work together even relatively simple tools can be used to kill predators. I was sent a video recently of a leopard suspected of killing people in India being killed by villagers. They had surrounded the animal in a field and simply closed in around it in an ever-tightening circle. At some point the leopard made a break for freedom, charging directly at a person in a style that is characteristic of leopards. That person was injured, but not badly, and the rest of the people present fell upon the cat and beat it to death with heavy sticks.

Lions in southern Tanzania are commonly killed by being speared, and Amy Dickman explained to me that the act of spearing a lion gains the hunter considerable prestige within the community. She went on to tell me that one specific role in the hunt involves giving away your spear and walking up unarmed to the bush where the lion is waiting in order to entice it out for others to spear. The reward for acting as bait is to have women dance for you. I am sure that must be quite a dance given the risk. This type of killing can at least be reasonably targeted towards a specific individual, but other forms of killing are far less precise. Snares, wire nooses that trap animals, account for the suffering and deaths of countless creatures. Lions may be targeted by snares set for them in retaliation for killing livestock or people, but far more commonly they end up as 'by-catch' in snares set to catch antelope and other large herbivores. Snares cause immense suffering as the captured animal writhes around trying to escape, resulting in deep and often fatal wounds. Foothold or leghold traps (also called 'gin traps') are also deployed and may catch lions, which are killed when the trapper returns. The most common method of retaliation, though, according to Dickman, is poisoning, where a carcass is laced and left for lions to find. In 2015 in Kenya, for example, Maasai herders laced a cow carcass with the insecticide carbosulfan after losing cattle to lion attacks. Three of the well-known 'Marsh Pride' of lions died in the incident.[48] Such poisonings are frequent events and it is common for many scavengers, including increasingly threatened vultures, to be poisoned alongside the targeted species. It is rare for these real threats to wildlife to gain any traction in the world's media and it is a great shame that the outrage generated over the Cecil killing does not seem to apply to the regular snaring, trapping, spearing and poisoning of lions across their range. Perhaps if it did, there might be more funding and resources directed to solving some of the real threats facing the world's great predators.

Informal – and what may also be illegal – methods of lion control have little regard for any sense of animal welfare and may have repercussions for many other species. Poisoning is especially indiscriminate, cruel and a major threat to a wide range of species. On the other hand, officially sanctioned interventions can be undertaken in ways that are targeted, precise and – with regard to animal welfare – considerably more ethical. If an animal is designated as a 'problem animal' then it can be dealt with under some form of Problem Animal Control (PAC) (with a PAC permit or similar officially issued permission). In practice, PAC most commonly equates to shooting the individual causing the problems.

Killing 'problem animals' is, to say the least, problematic. First, there are genuine conservation concerns. Killing an animal reduces the overall population, albeit only by one, but if the animal is of breeding age then future progeny are also removed. In some species, it is also the case that certain individuals may have a disproportionate role in the social structure and their removal could cause wider issues. This may sometimes be the case with dominant male lions and some elephants. Before these concerns become a problem, it is first necessary to find and kill the problem individual. In some cases, this might be straightforward. Some individuals may be clearly identifiable or have a predictable pattern of behaviour. Even if the animal is more elusive, many trackers have skills that are hard to believe unless you have seen them in action – and harder still to fathom when you have. However, it remains the case that animals other than the target animal may be misidentified and killed. The killing itself is also a difficult issue to navigate. A lion is a large animal and requires a combination of the right bullet hitting in the right place to ensure a quick and ethical kill. Even a very large bullet, placed very well and doing a huge amount of damage when it hits, may not instantly kill the animal. There are many videos online of lions that have been shot subsequently running off, leaping around and other horrors. These videos are usually accompanied by comments lamenting the shooter's incompetence and pointing out the welfare and ethical implications of the situation, but most videos simply reflect the fact that killing animals is not a clean and tidy process, and not something ever to be taken lightly. I have spoken with people who have shot problem lions in professional wildlife-management roles and my

lasting impression of those conversations is of the
respect they had for the animals and their desire to
ensure that the 'job' (as they saw it) was done ethically
and to the best of their abilities. Killing a lion through
an official 'death sentence' may be a difficult conflict
resolution to swallow, but it pays to reflect on what are
likely to be the alternative control methods employed,
quietly, indiscriminately and without official sanction.
And what may happen if someone doesn't make that
tricky decision.

Another frequent question raised around the issue of
problem animals is, 'Why do they have to kill them –
why don't they move them?' Such an intervention is
termed 'translocation' and involves darting the animal
with an anaesthetic to knock it out, transporting it
(most likely heavily tranquilised) to another location
and releasing it. Despite seeming more humane, and
aligning with what many would think are conservation
goals, it is not an intervention that is always supported
by most conservation biologists. There are many good
reasons for being hesitant about translocation as a
solution to problem carnivores – and because it is an
issue I will return to in almost every other chapter, it is
worth exploring these reasons a little now. Although I
will focus on lions, most of what can be said about them
applies equally to other species.

The problems with translocation start simply because
of the technological realities of darting an animal with
a 'capture gun'. Such guns fire a 'ballistic syringe' using
compressed air – they are basically a type of air rifle
firing a modified hypodermic syringe that discharges
its contents into the animal when it makes contact.

They can be very effective, but their range is limited – usually considered to be less than 30m (98ft). The closer the better really, because a syringe is not an ideal projectile to fling through the air and the precision of the set-up is not great, especially as range increases. Once you get beyond a certain range not only may you miss the animal, or hit it in the wrong place, but there may not be enough penetration for the drugs you administer to work properly. Throw in the effects of a little cross-wind and you really are looking at a technique that requires the operator to be close.[49] That is very far from ideal when dealing with a problem predator. Anaesthetics can work swiftly, but they are not instantaneous. What this means is that a problem animal, likely to be stressed and defensive as a consequence of being cornered, has been shot at close range with a syringe containing a drug with a lag time. The question now is, do you suppose a large animal with adrenaline pumping can cover the 10–30m (33–98ft) to the source of this new insult more rapidly than the anaesthetic takes effect? The footage I was sent from India of a leopard darted at 30m (98ft) showed that, in that case at least, the answer was a solid yes, and with more than enough left in the tank to maul the person who fired the syringe.

So, the approach is dangerous to people, but it can also be dangerous to animals. Pre-darting stress can be considerable and once the drugs take hold the animal may suffer adverse reactions enhanced by stress. Getting the correct dose is difficult and relies often on having a good idea of the animal's body weight. Animals can die under anaesthetic even when all the right precautions are taken. Once the animal is out, it must be moved to a new location and that transportation phase can be a complex mix of veterinary support (to keep the drugged animal alive), transport logistics and bureaucracy (before, during and after transport). Once these difficult and potentially very costly stages have been successfully navigated, it is necessary to release the animal somewhere suitable. This release site needs to work for the individual concerned, but it also needs to not cause issues for other members of the species already present at the site. Releasing a lion on to land that already contains lions will probably spark conflict between the interloper and the resident lions, potentially pushing the problem lion off prime habitat and closer to, say, livestock or human settlement. If the species is absent from the area then you must ask why, and also question the point of the translocation from a conservation perspective since the individual will not be able to breed. Overall, the problems very often outweigh the benefits and the cost may be prohibitive in any case. This is not to say that translocation can't work, and it is used in some cases, but it is not the go-to solution many believe it to be.

People are key
Human behaviour in its widest sense – and how our behaviour interacts with lion behaviour – is ultimately perhaps the most important factor in attacks. If we can change our behaviour, directing people away from risky behaviours, then lion attacks should reduce. How people perceive the risks of different behaviour versus the actual risk has an important role to play here. The risk-perception work undertaken by Kushnir and Packer in southeastern Tanzania, for example, showed that people

strikingly underestimated the risk of lion attacks around their homes, while overestimating the risk associated with farming or guarding crops. In other words, people felt they were more at risk than they actually would be doing activities away from the home, but felt far safer than they should at home doing mundane tasks (including going to the toilet). Kushnir and Packer equate this to the low perceived risk of a daily activity like driving compared to the higher perceived risk of the less mundane (but actually safer) activity of flying. Patterns of lion attacks established in different areas and our basic knowledge of human behaviour lead us inevitably to the conclusion that there won't be a single simple behavioural solution to reducing lion attacks in practice. However, developing an understanding of when, where and which people are most at risk – and ways to convert that academic knowledge into useful, practical behavioural change – will be crucial. Understanding the perception of risk, and where and why it may differ from reality, might well help deliver that impact, but could also help greatly in reducing fear and in turn the potential for pre-emptive lion killing. The message is clear: if we wish to conserve predators then we cannot ignore or devalue people.

CHAPTER THREE

Tigers

Nothing can really prepare you. Not even for just a photograph of what is left of a person after a large cat has killed and eaten most of them. The sight in real life must be unspeakably horrific. I have now seen many such images, ranging from small children to elderly people, and although the cold, terrible shock never really goes away, it inevitably subsides a little. But even as the horror fades through tragic repetition, I will always remember the first such photograph that I received from India in 2020. Gore is not something I seek out, but neither is it something that usually affects me too much. I can handle the sight of blood and guts, even when they are very much in the wrong place. So, when I opened that image, of a man who had been attacked by a tiger, it wasn't the mortal remains that were so viscerally, memorably and upsettingly vivid. What hit me about that image, and the subsequent ones I have seen, is how little a human being is reduced to when they become prey. A head, the hands, a section of spine and some flesh. This man, once like you and I – with a family, dreams, hopes and love – had become little more than offcuts in a butcher's shop, wrapped in a small piece of cloth ready to be transported to his family. It was, and is, truly awful and I can only imagine what his family must have gone through.

We should be shocked by such images. They are shocking. Every time a certain message thread pings with an alert on my phone I brace myself for another dose of the reality of what living with predators can be like. That message thread originates in India and largely concerns the subject of this chapter, the tiger. Of all the animals that make us prey, tigers have perhaps the fiercest reputation, but not that long ago I would have assumed that this reputation was one built in the past, on the legacy of Jim Corbett (who we met in Chapter One and will hear more about shortly) and barely relevant in the modern world. Tigers, after all, are the very icon of an 'endangered species' and the impression that there are scarcely more than a handful left, clinging on in national parks and protected areas, is one that is hard to ignore if you engage with wildlife stories in the media. With so few tigers left, and those that remain living in areas where there are few people, then the potential to be attacked by one is surely extremely slim? Well, let's just say that the reality of modern-day tiger–human interactions in India (and to a lesser extent Nepal) is a little more complex than you might think.

The tiger: a biography

Tigers are close relatives of lions. Both species belong in the genus *Panthera*, which also includes the other big cats, leopards, jaguars and – recently added – snow leopards. Tigers and lions can actually interbreed, such is their relative zoological closeness. A male lion mating with a female tiger can produce a hybrid offspring known as a liger, while the opposite mating combination produces a tigon. Other *Panthera* hybrids exist in captivity, including jaggers, jaglions, jagupards, lipards and leopons. It is often said that these hybrids are sterile, but that is not entirely true. In some cases, the female hybrids may be able to reproduce successfully – as demonstrated in 2012 when the Novosibirsk Zoo in Russia reported the birth of a 'liliger', the offspring of a female liger and a male lion.

Despite being closely related, tigers differ from lions in a number of key ways, most obviously their orange colour and black vertical stripes. In addition, tigers are larger than lions, although there is a considerable variation in size across their range and between sexes, with female tigers being markedly smaller than males – 65–165kg (143–364lb) compared with 90–300kg (198–661lb). The largest tiger ever is something of a disputed category. A Bengal tiger weighing 389kg (858lb) shot in India in 1967 is generally accepted to have been the largest-known wild tiger, but he had a very full belly of buffalo from the night before. A captive Siberian tiger called Jaipur weighed 423kg (933lb), but one can safely assume that, being a captive animal, he wasn't quite the lean, mean, muscle-bound machine that a wild tiger would be.[1] Even with some disputes, and the influence of captivity and a full belly, it seems realistic to suggest that some wild tigers, especially in the past when larger tigers were more common, would have tipped the scales at over 350kg (772lb) – and that is a lot for a cat. Some lions in captivity might approach that size, and one wild lion (in fact a 'man-eater') was shot in South Africa in 1936 that weighed 313kg (690lb),[2] but these are exceptional individuals. Overall, a good-sized male tiger is likely to be bigger than the vast majority of lions.

Finally – and an important point when considering feeding biology and habits – tigers are solitary animals and do not form prides or other social groups. In this regard they are similar to other big cats, with lions being the social exception among *Panthera*.

Tigers are most associated with India – and this is certainly their stronghold in the present day – but historically their range went far beyond the subcontinent. At one time, tigers could be found from eastern Turkey across Asia to the Sea of Japan, and south through parts of Russia, central, south and South East Asia, and east into Indonesia. This is not the case now, although they can still be found in a number of countries throughout Asia. Of the 14 countries identified as Tiger Range Countries (TRCs), most tigers (ca. 70 per cent) are found in India, but they also roam free in Bangladesh, Bhutan, Cambodia, China, Indonesia, Laos, Malaysia, Myanmar, Nepal, North Korea (although the current status of tigers in the Korean peninsula is unknown), Russia, Thailand and Vietnam.

The existence of tigers in multiple countries spread out across Asia gives a false impression of the extensiveness of current tiger range. Within each of the TRCs tigers are highly constrained, often existing in fragmented populations within protected areas. Overall, their current range is a pitiful scrap (perhaps just 6 per cent or less) of their formerly impressive geographical sweep and there are some doubts as to the status of tigers in some parts of what is considered to be their current range, notably the Korean peninsula. Further complicating the geographical mix is the fact that, linked to their distribution, there are multiple subspecies of tigers.

Populations of a species that, while clearly belonging to the same species, have some clear and definable differences may in some cases be considered to be subspecies. As you might imagine, defining subspecies – and getting biologists to agree on their status – can be challenging. Subspecies may eventually end up being recognised as full species – as happened recently with giraffes, which are now considered to be four species, rather than a single species with four subspecies. The number of tiger subspecies turns out to be rather difficult to pin down and opinion varies, sometimes fiercely so. Historically, the approach was to recognise nine subspecies that differ in size, coloration, stripe pattern and hair length. Those nine are the Amur (or Siberian), Indian (or Bengal), South China, Malayan, Indo-Chinese and Sumatran, plus the now-extinct Bali, Caspian and Javan tigers.[3] All these subspecies may seem a little excessive when you look at the animals themselves, but first impressions can be misleading. Traditionally, subspecies were decided upon through the minute examination of morphological and other characteristics, comparing individuals with those typically living in the same population with members of the species living elsewhere. This approach is still valid, and these days we can throw sophisticated measuring and statistical techniques into the mix to give us even more resolution. Of course, due to our understanding of species, or populations, and especially the flow of genes between populations, we can also take a molecular approach to the subspecies question. Genetic analyses allow us to look at more fundamental differences and similarities, and to examine what may have happened historically as species spread. When tigers were analysed using multiple lines of evidence, the results were disappointing for fans of splitting species into subspecies. Two subspecies were proposed: a northern group (the Siberian and extinct Caspian tigers) and a southern group (all the rest).[4] The IUCN Felid specialist group in 2017 had only those living in the Sunda Islands as a separate subspecies from all the rest. The issue appears to have been resolved in 2018 with whole-genome sequencing from 32 specimens identifying six 'statistically robust' subspecies.[5]

Subspecies identification might seem like an academic exercise, especially in the case of tigers where none of the subspecies is likely to be confused with any other species. These animals are, after all, unmistakably tigers. However, subspecies classification can be important for conservation. If we want to conserve as much biodiversity as possible – and as much genetic diversity as we can – it is important to recognise where species have naturally diverged into identifiable 'evolutionary significant units'. Subspecies will probably have traits that adapt them to live in the conditions prevalent in the area in which they evolved, but there is also some gene flow between subspecies. Understanding the evolutionary history of species and subspecies can inform our conservation approaches, including captive breeding and species reintroduction.[6]

Tigers in trouble

As well as a hugely contracted geographical range, tigers have seen a massive drop in population, especially over the past century or so. This is entirely our fault. As we have already seen with lions – and will see with most of the predators in this book – it is the classic anthropogenic 'trident': we have hunted and persecuted tigers across their range; we have depleted their prey; and we have reduced their habitat. There were perhaps 100,000 tigers a century ago and the current population is disputed (of course) but stands at around 3,900 (and rising).[7] As we have seen already with lions, though, counting wild animals is very far from straightforward. Tigers are elusive, secretive animals, well camouflaged in their forested habitats. Furthermore, a long history of persecution and habitat loss has tended to push existing tigers into more remote, less accessible areas away from people. None of these characteristics makes surveying them easy, but resources are directed at studying them in the field, due to their status as one of the world's 'iconic' species. That means that we would be wrong to hold out much hope of some pocket of undiscovered habitat loaded with thousands of uncounted tigers.

If you're upset that we have lost about 95 per cent of tiger habitat and individuals over the past 100 years or so then you should be. It is a terrible indication of our slapdash approach to conserving the natural world that arguably the most recognisable of our predators has dwindled to such an extent. However, the past decade provides some hope. Tiger populations are either stable or increasing in Nepal, Bhutan, Russia, China and, best of all, India.[8] This resurgence has come as a consequence of a remarkable conservation pledge made in 2010 by leaders of 13 countries to double tiger numbers by 2022. TX2, as the initiative is known, has seen a huge global effort dedicated to the species, with habitats and corridors (connections between areas) being restored, protection being granted and, crucially, local communities and politicians being brought on board. Tigers are still under threat and efforts must continue, but there is some cause for hope.

Human predation

Tiger recovery means more tigers and, over time, a wider geographical range. That is great news, but in all but the most sparsely populated areas of tiger range an increase in tiger numbers will inevitably lead to more interaction with humans. Nepal was the first of the tiger range states to double its tiger population. In 2009 there were an estimated 121 tigers and in 2018 this had risen to 235, agonisingly close to the TX2 target some four years before the 2022 goal. In 2021, when I spoke to Sam Helle, a tiger researcher from Minnesota working in Nepal, she confirmed that Nepal had indeed made it over that doubling threshold and that tigers continue to increase in the country.[9] The Bardiya National Park, in the lowland region to the south-west of the country, fared particularly well, with tigers increasing nearly fivefold, from 18 in 2009 to 87 in 2018. This population increase occurred without any reports of fatal human–tiger conflict, despite 12 people being killed and four injured between 1994 and 2007.[10] However, that all changed across the last half of 2020 and the first quarter of 2021.

During that nine-month period, 10 people were killed by tigers within the National Park and each death was a human tragedy of untold misery. Let's imagine just for a moment that the same number of people died in the UK from some new hobby or craze. Headlines and outrage would turn rapidly to bans and legislation long before the death toll made it into double figures. If the same number of people had died as a consequence of an animal? Well, front pages would surely be calling for culling, control and, I have no doubt, eradication. And yet, I struggled to find mention of these deaths in any media outside Nepal and India. The only mention of the Bardiya National Park deaths in the UK press I could find was a short piece in *The Times* in April 2021, largely bemoaning the fact that the deaths were delaying the rhino census. The piece did at least name the 10th victim, Policeram Tharu. If our conservation goals are to increase predator numbers then we have to realise that human deaths may be tied up in that goal. Rather than wave them aside as an inevitable consequence of increased predators, a dismissal that is so much easier when the people affected are elsewhere, we must try to understand how and why these attacks happen. In the case of the Bardiya National Park attack cluster, it isn't as simple as more tigers resulting in more attacks. We shall return to Nepal shortly, but first let's travel south into India and examine our historical relationship with tigers.

Humans have coexisted with tigers for millennia, but it would be wrong to assume that this coexistence was peaceful. In the early 1900s it has been estimated that 1,000 people a year were killed by tigers in India, but this figure should be treated with caution. In the present day, with modern record keeping and data collation, we have very little idea how many people are killed by tigers in India, so the round number of 1,000 feels like exactly what it is, a guess. In this case, I think it is a low guess. The most comprehensive study of tiger-related attacks and deaths I could find suggests that 373,000 people died between 1800 and 2009, which is close to 1,800 people a year over that period, and in total is more than the current population of Iceland. Of course, deaths will not have remained constant over that period, with tiger numbers declining and humans increasingly dominating the landscape; the pattern of conflict will also likely have changed. All of this makes it difficult to estimate human deaths from tigers.

Why might 1,000 deaths a year be an underestimate? Well, not all tigers end up eating people, but some do, and some of these tigers do so with considerable prowess. A single tiger in Bardiya probably accounted for half of the victims, at least according to local reports.[11] However, in the grand scheme of tiger attacks, five deaths attributable to a single animal is a low number. The Champawat tigress purportedly killed 436 people in the last part of the nineteenth and early part of the twentieth centuries – around 200 in Nepal and the remainder in the Kumaon area. This number is staggering and at first glance seems unbelievable. That a single wild animal could, over time, kill that many people seems so alien initially that it is hard to comprehend. However, a fully grown tiger will kill and eat at least one large, deer-sized animal every week. The Champawat tigress killed and ate mainly woman and children, who generally weigh less than a chital, a species of deer commonly hunted by tigers. Humans are a far easier target than a chital, especially if they are alone and unarmed, as would have been the case for many victims.

Even if we assume a kill-rate of one per week (which is likely to be an underestimate) then 436 people represent nine years of predation, a period of time more or less equal to the period over which the Champawat tigress was known to be active. Once we distance ourselves from the human suffering, treat the tiger as a predator and humans as prey, then these large historical numbers suddenly do not seem so unbelievable.

The period over which the tigress was active, especially during the latter years, became years of fear for those living in the area. The tigress would travel as far as 32km (20 miles) between villages and between attacks, so that people never knew where next to expect her. Her hunting territory grew, as did the number of villages and potential victims. As we saw with lions, certain types of human behaviour made some people far more likely to be attacked than others. Women and children out in the day collecting firewood, feed for livestock or flowers were common targets. Great efforts were made to kill the tigress and, during the early years, the Nepalese army even got involved. This was good news for the people of Nepal, but bad news ultimately for those over the border in Kumaon. The army, unable to capture or kill her, drove her across the border into India, thus relocating rather than solving the problem (which harks back to the discussion in Chapter Two about the problems of predator translocation). It would be 1907 before the tigress was killed, by a remarkable Englishman called Jim Corbett.

Corbett was a hunter, but to sum him up in a word so loaded with negativity in the modern world, and so simplistic, is to do a huge disservice to the man.

A colonel in the British Army, Corbett had grown up in Kumaon and had immersed himself in its forests and the wildlife living within them. As a skilled tracker and naturalist, he became famous for tracking down, and killing, tigers and leopards that had started to hunt and kill people. He wrote up these accounts in best-selling books including *Man-Eaters of Kumaon*, the depressingly necessary sequel *More Man-Eaters of Kumaon*, *The Temple Tiger* and *The Man-Eating Leopard of Rudraprayag*. While Corbett became known initially for killing big cats, he was very far from the bloodthirsty white-hunter stereotype. He wrote with compassion and sensitivity towards the people affected, and he had a knowledge, respect and love for the forest and its inhabitants that shines through in his prose. He dedicates one of his books, *My India* published in 1952, to '... the poor of India', saying that, 'It is of these people, who are admittedly poor, and who are often described as "India's starving millions", among whom I have lived and whom I love, that I shall endeavour to tell in the pages of this book.' Writing 70 years ago, Corbett understood far more than most people in today's world that it is the poor who shoulder the heaviest burden when it comes to predators.

Corbett seems in many ways to have been a reluctant tiger hunter, often taking on the mantle simply because he could. His guilt when a mistake caused a tiger to escape and go on to kill again is clear, but he always returned to the task, his tracking and hunting experience furnishing him with the very particular set of skills needed to solve the problem – and solve them he did. After dispatching the Champawat tigress, Corbett went on to kill two prodigious human-predating leopards and a number of notorious tigers. In 1938, Corbett killed his last tiger, the so-called Thak man-eater, a tigress that had killed four people in Kumaon.

In later life, Corbett also used his forest skills to record tiger behaviour on cine-camera. He went on to become a champion for tigers and helped to establish a national park in Kumaon that was named after him in 1957, two years after his death. Corbett loved tigers and the natural world, but it was very much his understanding that it was unacceptable to expect people to live alongside tigers (and leopards) with a history of hunting down and killing people. We will compare this with the modern-day approach to man-eaters in India shortly.

As an aside, Corbett crops up as a charming footnote in the story of the British royal family. He and his sister retired to Kenya in the late 1940s, and in 1952 he acted as bodyguard (and presumably rather excellent field guide) for Princess Elizabeth when she visited the famous Treetops Hotel, built into the branches of a huge fig tree. It was during her stay there that her father George VI died and Corbett wrote in the visitor book, 'For the first time in the history of the world, a young girl climbed into a tree one day a princess, and after having what she described as her most thrilling experience, she climbed down from the tree the next day a Queen – God bless her.'[12] As a visitor-book entry it certainly beats, 'Great stay, recommended.'

The Bardiya forest attacks
Tales from Corbett's time in India give an indication of the extent to which tigers were a problem to people at

that time. In Kumaon alone, Corbett killed cats that he estimated accounted for 1,200 people. Notably, the first of the man-eaters he killed, the Champawat tigress, was from Nepal. The present-day tiger attacks in the Bardiya National Park show that history can repeat itself, but the situation in modern-day Nepal is very different from that of Corbett's era. For a start, tigers are far fewer in number, but increasing, and far more constrained in terms of habitat. In contrast, people are more numerous and more widespread across the landscape. Habitat encroachment and hunting means that natural prey is depleted. The wider landscape has become progressively more human-dominated, with tiger habitat reduced largely to fragments of protected areas surrounded by – and in some cases connected by – areas where humans and their livestock are common features. This combination of factors and landscape changes can certainly account for the rise in tiger attacks. As Ashim Thapa, information officer at Bardiya National Park, said in an interview given to The Third Pole (an online news and information platform devoted to the Himalayas) 'not every tiger is a man-eater. But as the numbers of tigers rise, there is an increased risk of attacks that can happen when old and weak tigers roam around in search of new territory and prey.'[13] However, there are other factors at play that may be specifically influencing the rise in tiger attacks in Bardiya.

Deaths were occurring mainly when villagers entered forests to collect firewood, fodder and timber. According to a survey of the people in the area, the primary reason people were being driven to enter the forest was poverty.

You will remember that poverty was also a key factor driving lion attacks in Tanzania. Impoverished people, entering the forest to gather very basic resources, end up being eaten by tigers and yet are the same people you will often see condemned on social media as 'asking for it' by 'being in the tiger's territory' (I'm paraphrasing because many such comments aren't worthy of verbatim repetition). Poverty is nothing new around the Bardiya National Park, but Covid-19 increased the pressure it causes because the absence of tourists translated to an absence of income. On top of that, to protect people from tigers, the park authorities put a halt to walking safaris, blocking off another potential source of income for local people who act as guides. So, poverty is exacerbated by the response to the global pandemic and – coupled with a successful conservation programme that has resulted in greatly increased tiger numbers – develops the conditions necessary to see dramatic rises in tiger attacks.

Another factor that folds into this 'perfect storm' is water, more specifically the Geruwa river. The river forms the western boundary of the National Park, creating forested habitat alongside its bank while providing water for wildlife and for people. More than a decade of mining gravel and rocks near the headwaters has resulted in a decline in water level and changes in the river's course. This, along with a recent dry-season drought, has led to water levels low enough that animals, including tigers, are seeking water in villages. Just as desperate people seek resources in the forest, where they would rather not be, so too desperate animals are forced into human settlements. Low water levels in the river also make it easier for people to cross and enter into the forest, and for animals to leave. Poverty, the pandemic, gravel extraction, climate change and positive conservation action for tigers have all combined, with poor outcomes for local people and ultimately for wildlife.

Complex problems, with multiple factors, are rarely solved with simple solutions, although in this case a 'simple' solution does present itself. Removing all tigers from the area is clearly one way to prevent further attacks. History tells us that predator eradication is perfectly possible, as does the 95 per cent of tigerless former tiger range. Complete removal of tigers from the park, while feasible, is not a viable solution, although any suggestion that it might occur would doubtless attract far more global media attention that the 10 people who lost their lives. Removal of the problem animals is a better alternative, but as we have already seen with lions, it is far from straightforward to do even if there is political will and local authority to do so. The ramifications of lethal control, especially with the media savviness and political clout of animal rights organisations around the world, go far beyond the boundaries of any national park. International reputation, direct funding for conservation and other aid, as well as tourist appeal, may all be negatively affected by such direct action. When Botswana announced that it was to reopen some very limited hunting of its 130,000 and rising elephant population to provide financial incentives for local communities affected by crop raiding, and in some cases human deaths, there were open calls from groups to boycott Botswana as a tourist destination. For Bardiya, a region relying heavily on tourists coming to view tigers, such attention could be fatal.

With lethal control effectively taken off the table, local authorities have to look to manage tiger–people interactions. The easiest way to do this initially is by controlling human behaviour – or trying to. In January 2021, park authorities restricted traffic on a highway section that runs through the park, banning two-wheeled vehicles and pedestrians. It doesn't take an economics genius to work out who is most likely to be on foot or a two-wheeled vehicle, so such restrictions preferentially disadvantaged the poor while simultaneously attempting to protect them. Some buffer zones around the park, including a community forest, have also been shut off and park officials have banned people from entering the forest alone to collect fodder and wood. Banning people from doing something is very different from preventing them from doing it. It is little surprise that the attacks continued despite these measures.

If solving forest-access problems against a backdrop of poverty might be difficult, then solving the bigger issue of declining water levels in the Geruwa is proving nearly impossible. Since 2015, 152 artificial ponds and solar pumps have been installed by the National Park but, according to Rabin Kadariya, the head of NGO the National Trust for Nature Conservation (NTNC) in Bardiya, 'it hasn't been proven sufficient to address the ongoing issue. The water crisis seems to be getting worse.' Preventing further upstream development would also be a good move, but Kadariya points out that 'with the ongoing construction of the Rani Jamara

Kulariya irrigation canal and the multi-basin irrigation and hydropower project at Babai [upstream of the National Park], the crisis will be aggravated in the future'.[14]

So, controlling tigers is fraught with issues, preventing tiger–people interactions is tricky, and solving the underlying water crisis that exacerbates the problem seems to be impossible. If all of this seems depressing, then welcome to the world of real-life conservation. I said earlier that complex problems, with multiple factors, are rarely solved with simple solutions, but actually one way to solve the problem is both simple and achievable. People get attacked by tigers because they have to enter the forest where tigers live to collect resources, and the poorer people are the more likely this is to happen. Likewise, people walk or cycle along forest highways when they are too poor to afford a covered vehicle. So, make people richer and you solve the problem in a stroke. If we can find ways to make sure that people are not dependent on the forest in ways that put them in danger and if we can provide them with more lucrative opportunities that involve them in conservation, rather than exclude them, we might just solve the problem. Remarkably, elsewhere in Nepal we can see community conservation in action in a model that has much to teach us for conservation in general – and predator conservation in particular.

A major component of the general success story of tigers in Nepal is the community forestry programme. Starting in the 1970s, and developing over the past few decades, the community forestry programme formally decentralises control of forest management, putting local people in charge of their resources. The forests were formerly degraded through overuse, but as a consequence of local management they have recovered. By 2009, a third of Nepal's population was involved with the community forestry programme.[15] Collectively, these communities were actively managing a quarter of the forests of Nepal. By being able to manage their resources and benefit directly from them, local communities were empowered to manage forests sustainably and to develop new sources of income. As the forests recovered, so too did prey species. A combination of habitat and prey restoration, together with local protection, has been a major factor in the recovery of tigers. This approach is the direct opposite of the 'fortress conservation' model, where local people are excluded and ignored. By recognising and encouraging local communities' rights to use their resources, and their ability to do so in a sustainable way if given the right tools and support, Nepal provides a model for community-empowered conservation.[16] Taking care of local people's needs led to enhancement of the environment and, in time, a conservation-focused management agenda that has worked for tigers and people. We should of course be cautious, and not think that conservation benefits for certain species automatically flow from such approaches. We should also be mindful that community management can still lead to over-exploitation of resources and inequality of wealth division. Nonetheless, Nepal's community forest-management programme clearly gives us pause for thought and cause for hope. It is certainly notable that where people are losing out in Nepal to tigers the costs are centred on a national park, with centralised rather than community management.

Back to India
Back over the border in Kumaon, India, things may have changed since Corbett's day – and the time of the Champawat tigress eating more than 400 people – but tigers remain a threat in this area. Sunderkhal, a village close to the edge of the Jim Corbett National Park, was the centre of activity for a tiger that killed seven men and women over several months from September 2010. The animal was shot by the authorities on the outskirts of the village in February 2011, following a period of fear for local people, who laid a 'virtual siege' at the offices of the National Park in January 2011. A report from the area some four years later reports that 'fear refuses to leave the village'.[17]

Getting definitive information on the scale of tiger attacks across India right now is difficult, but one man who has tried to compile a database is Rajeev Mathew. Just two hours before typing this paragraph I received a message from Mathew of a press report in a local paper in India. The translation reads as follows:

The body of an elderly man has been found in the core zone of Tadoba Andhari Tiger Reserve [in Chandrapur District, Maharashtra state, central India] at 8 a.m. on Saturday, June 12. The deceased, Bharat Ramaji Bavane (65), a resident of Mudholi, was missing since 3 p.m. on Friday. With this incident, the death toll in tiger attacks since January has gone up to 22. The Forest Department has appealed to the villagers not to go into the forest, but

still people do not pay attention to this and lose their lives
[similar to the situation in Nepal]. A resident of Mudholi
village went to cut bamboo in the nearby forest on Friday.
At 9 p.m., information was given to the Forest Department
Officer (Wildlife) office of Moharli. As soon as the
information was received, a team of the Forest Department
started searching. But they did not have success. This
morning the team again started the search in the forest,
then at 8 a.m. in the Ambegad fixed area of Moharli Wildlife
Zone, his body was seen about 85m from the boundary of
the buffer zone. Bhadravati police reached the spot and sent
the remains to the rural hospital for post-mortem. With
today's incident, the death toll for tiger attacks since January
has gone up to 22.

Twenty-two people dead in six months from tiger
attacks, in one district. It sounds like Jim Corbett's days
are returning.

There is no central database compiling details of
animal attacks and Mathew has to piece together his
dataset from local press and media reports, as well as
information he gets directly from different sources
around India. It is not an easy process and neither is it a
complete record. Some victims may never be found,
while other deaths go unreported or may be missed in
the huge number of different local news publications
produced in multiple languages throughout India.
Mathew has helped me out on a couple of radio projects
and he has been kind enough to share some of his
findings with me. They make for distressing reading.
According to Mathew, 'Last year – January to December
2020 – there were 80 attacks by 66 different tigers across

India. These are numbers I was able to confirm. Many states where the conflicts are very high are not reporting kills, or local media is not accessible. As a very base average, I reckon last year at least 200 people lost their lives to tigers.'

He goes on to reference the area where 22 people have died already, but Mathew suggests that number is probably higher: 'This year so far I have registered 50 deaths and injuries almost all of them (barring five) are from Central India, especially around Tadoba Tiger Reserve [Tadoba-Andari Tiger Reserve]. Even these are not complete.'

When we bear in mind that estimates 100 years ago – when tigers were perhaps more than 25 times more numerous – were of 1,000 deaths a year, we can conclude that estimates in the past may have been underestimates, as I suggested earlier, or that tigers are more likely to attack people today than in the past, or both. Mathew suggests that range expansion of tigers in recent years, coupled with long, multi-generational absences of tigers, may be a factor in modern-day attacks. According to him, 'In many cases we are seeing tigers [in some areas] after a lapse of anywhere between 80 and 200 years. People have forgotten how to move about in forests and take precautions. That leads to fatalities.'

Mathew also thinks that the absence of hunting of tigers by people, outlawed in the 1970s and essential for their recovery over the past few decades, also resulted in tigers now having no fear of humans. So, we have a predator with no reason to fear us and people who are no longer used to functioning in environments with predators.

According to Mathew, the people most likely to be attacked are, you will be unsurprised to learn, 'the poor, especially the landless, and the old'. As in Nepal, we can see unintended negative consequences from Covid-19 restrictions playing out in India. Mathew outlined to me how 'lockdowns have forced people working as daily wage earners, unskilled and doing other odd jobs to go back to their native lands. This has thrown a great burden on the forest and these people are competing with wild animals for the same resources – forest produce. A small percentage of people in the forest is tolerable, but when there are a large number, it causes immense disturbance. Many of the people killed are because of that person being at the wrong place at the wrong time.' It's the same pattern we saw in Nepal and effectively the same pattern for lion attacks in southern Tanzania.

Understanding how tigers hunt is an important part of understanding how and why people fall prey to them. When a large cat engages with a person in a predatory attack, the interaction is not frenzied. I asked Mathew how tigers normally attack and his answer was 'with precision'. A typical tiger attack will kill the target, whether a human or a deer, with a simple bite through the chest or neck. Tigers have the longest canines of any big cat, reaching more than 7cm (2.7in) in length. These four dagger-like teeth are separated with smaller incisors at the front of the mouth and have a significant gap behind them before the premolars start. The wide spacing of the canines gives them a formidable 'gape'. Powerful jaw muscles in the head allow a tiger's bite to drive those canines through the flesh and bone of the neck or chest until they meet. The gap behind the front teeth makes it easier to hang on to prey as it struggles, although the reality is that prey will not struggle for long. Larger predators, taking large prey animals, risk injury if the prey is left alive and kicking; the goal is to subdue and kill rapidly.

The final moments for a human falling prey to a tiger are the same as for a deer, as are the moments leading up to killing. Mathew explained to me that 'tigers hunt us in the same way as any prey they normally hunt. Their feet are cushioned with pads, and their foot placement is very delicate and one cannot hear them if they are stalking prey. They move extremely quietly and will take an age to cover any ground, if in the open, otherwise they move quite rapidly, yet stealthily.' If a tiger intends to make you prey, you are unlikely to know anything about it until the very last moment. As well as being stealthy, tigers are strategists. Mathew, who has plenty of experience of tigers in forests, says that 'they also study their quarry. They understand humans very easily since we, like them, are creatures of habit. Usually they mark a person, man or woman, and will stalk the person till they come within striking range.'

Some of the activities people undertake in forests make them especially vulnerable. Mathew's studies show that gathering forest produce was the main activity people who were attacked were undertaking. Mahua (*Madhuca indica* and *M. longifolia*) flowers are in great demand commercially, as are the fruit they go on to produce, while tendu leaf (*Diospyros melanoxylon*) is collected and dried to make cigars known as *beedi*. In the north of India the harvesting of sugar cane is also a risky activity as many tigers (and leopards) live and breed in cane fields, which are left undisturbed for almost a year. Meanwhile, cattle grazing can bring herders into direct conflict with tigers. Cattle in India are often turned loose and allowed to roam semi-feral in the forests. When tigers attack, herders may intervene to protect their charges and fall prey themselves. Confining cattle is a mitigation technique we have met before in Africa (*bomas*), but cattle *kraaled* close to, or even within, dwellings can lead to tigers raiding the dwellings and killing people.

Jim Corbett emphasised throughout his accounts that the tigers he shot were old or injured animals. A post-mortem on the Champawat tigress, for example, revealed that her upper right canine was broken in half while her lower right canine was snapped off at the jawbone. Corbett determined that the injury had been caused by a bullet. The Thak tigress had also been shot, carrying two old wounds – one of which, in her shoulder, had become infected. Corbett writes, in the foreword to *Man-Eaters of Kumaon*, 'The wound that has caused a particular tiger to take to man-eating might be the result of a carelessly fired shot and failure to follow up and recover the wounded animal, or be the result of the tiger having lost his temper while killing a porcupine.' His clear belief was that tigers had to be 'infirm' in some way to resort to taking human prey. This may well have been the case in Corbett's time, when the high pressures on tiger habitat and prey availability we see now were less severe. In modern-day India it seems that this 'old and infirm' stereotype is not so prevalent.

Mathew tells me that in modern-day India, as in Corbett's time, most old tigers do not turn to human prey although 'there are several cases reported, especially from the south, the Western Ghat complex and from the northern regions, especially around Corbett National Park, Dudhwa and others'. What is interesting, though, and what seems to be an almost complete reversal of the Corbett pattern, is that, according to Mathew, 'In central India the culprits are usually the young animals. Most grow out of it once they establish territories.' Such an apparently large change in behaviour demands an explanation and Mathew has spent some time trying to account for this pattern. His hypothesis links directly to the success of tiger conservation in India. He says, 'The population of tigers is exploding, especially in central India. I can say this with some certainty since I work in the landscape quite extensively. There are more than a hundred cubs in the age group of between eight and 27 months. Younger cubs are present, but are not taken into account.' Mathew describes a landscape with a relative abundance of tigers and a large number of younger animals. He has also often noted to me that tigresses might be seen with four or five cubs at heel now, whereas in the past two was a more common number. Craig Packer, when I interviewed him about lion breeding and population recovery, said that 'lions breed like cockroaches' and I have heard others say this for big cats in general, especially lions and tigers in captivity. With active conservation efforts to protect them, no natural or human predators, and with habitat management increasing their prey base, tigers are doing well. The large number of sub-adult animals Mathew reports indicates that many of the cubs he is seeing trailing after their mothers are making it to a size that becomes a problem. A young tiger is still a tiger and, still growing, it can consume 20 per cent of its own body weight in a single sitting. Mathew notes, 'Bandhavgarh National Park has 44 cubs! The area of the park is 1,536 square kilometres. I also reckon on 100-plus cubs in the district of Chandrapur in Maharashtra. One can imagine what the situation is going to be like in less than a year from now.'

While great for recovering tiger numbers, this population bubble of younger animals building up rapidly in a protected area, if true, could be storing up trouble. To understand why, we have to understand tiger ecology and behaviour. Tigers are solitary animals and both sexes are territorial in different ways. Female tigers have territories that are generally exclusive to one female, although there are records of female tigers overlapping or even sharing territories. In general, though, one female will occupy a piece of home range that it calls its own and that it delineates to other tigers, and possibly other animals too, with scent marking. Male tigers, on the other hand, have much larger territories that overlap with one or multiple females. Male tigers generally do not overlap with other males. The territory sizes differ massively depending on where they are because to a large extent they will be determined by prey availability, topography, natural barriers and so on. As a rule of thumb, female territories are somewhere around 10–50km^2 (4–19 square miles) and those of males around 25–200km^2 (10–77 square miles), although they can extend up to 1,000km^2 (386 square miles) in parts of Russia.[18] The main driver for males is females, which means that male territory size is driven by female territory size and how many females the male

is able to defend from his rivals. Given the vast areas
involved, male tigers need to cover a lot of ground, a
feature we saw in the account of the Champawat tiger.

These territories, or home ranges, tend to be stable
over multiple years, and carving out a territory of your
own is tough for a younger animal. Territories are won
by force and the simple reality of animal combat is that
the larger individual usually wins. Age can play a role
and, once they are fully grown, younger animals may be
quicker, but they are not necessarily stronger and they
are certainly less experienced. This means that when
tigers leave their mothers at 24 months of age they are
unlikely to slip straight into a cushy territory like some
well-connected intern landing a job after a summer of
work experience. A two-year-old tiger is no match for
an experienced territory holder and these younger
individuals will have to spend some time wandering,
pushing boundaries and losing fights before they can
establish themselves. It is this feature of tiger ecology
that clashes with the modern landscape. As Mathew
explained, the younger tigers 'are usually pushed out of
the protected forest areas and quickly find themselves in
human habitations. They may try to kill cattle and they
will kill the people residing in the houses and huts
nearby, especially if the owner of the cattle comes out to
check on the commotion.' In a less human-dominated
landscape these younger tigers would serve their
apprenticeship deep in the forests, but with those
remaining forests occupied by territory-holding tigers,
they find themselves in contact with people. As Mathew
says, with a large degree of understatement, 'A country
that saw an almost extinction of tigers is suddenly being

seized with an overabundance of these animals – and scientists, managers, politicians and also the public at large do not know how to react. That is a major management problem as I see it.'

The Sundarbans

Mathew's observations focus on areas around central India and in the north-east of the country, but there is a region in this area that was, and still is, notorious for tiger attacks: the Sundarbans. This region, straddling the border between India and Bangladesh, is a vast area of mangrove forests, agriculture, mudflats and islands formed in the huge delta laid down where the Ganges, Brahmaputra and Meghna rivers meet. The Sundarbans are a rich habitat, recognised by UNESCO World Heritage status, and with multiple protected areas where wildlife thrives. Tigers in general have a higher than usual affinity than other cats for water, and in the Sundarbans this attribute comes into its own. Tigers are able to move around the mangrove forests, swimming between islands in search of prey like chital deer, wild boar and, sometimes in large numbers, the people that also make use of mangrove forests.

Perhaps because of the geography and topography of the region (which arguably suit tigers far more than people), the relatively high density of tigers, the low density of natural prey, the relatively high human population (more than 3.5 million people live in the Bangladesh Sundarbans) and the fact that around a third of people rely directly or indirectly on forest resources, tigers in the Sundarbans have long been known for killing and eating people. Just in the Bangladesh

Sundarbans, 1,396 deaths were recorded in the 63 years between 1935 and 2006, with an average of 22 per year.[19] As before, we can safely assume that this number is most likely an underestimate. People undertaking illegal activities in the forest or those who are attacked and die some time later may not be reported by people they are with, or may not be recorded as a tiger-related death by the authorities. The number of tiger-related deaths that may be missed in official figures has been calculated at 33 per cent for this region,[20] which takes the number of people killed annually to around 30, or one every 12 days. We must also bear in mind that these numbers apply to an area of no more than 6,000km^2 (2,317 square miles), or about one-third the size of Wales.

An analysis of tiger attacks in the Sundarbans region showed a wide variation in deaths per year from 1935 onwards, sometimes fewer than 10 and at other times approaching 100. Overall, though, the number of deaths per year has remained stable in the sense that, on average, as many people are killed now as in 1935. The analysis also allowed some insights to be gained in terms of who is being killed and which tigers are doing the killing. Most attacks happen in the forest areas as opposed to around human habitation (the ratio being around 15:1) and more attacks occur during the months of March–June, when the climate is drier and people move into the forests to collect wood, honey and mangrove palms. Mangrove palm, also called nipa palm, is widely used as a roof-thatching material, and for basketry and other weaving crafts. As you can imagine, it isn't the wealthy urban elite who find themselves entering the forest to forage for such

materials. When the rains begin, fewer people venture
into the forest, although some of those who do may be
there illegally and attacks on them may be less likely to
be reported. Most attacks happen during the day, but
this is largely because people don't tend to venture into
the forest at night. Of those who enter the forest, some
people seem more likely to be attacked than others, with
honey collectors, known as mawalis or moulis, being
the most vulnerable. This is probably because they must
cover a large amount of ground and – although they
work in groups – they tend to be quite dispersed and
communicate with others though vocalisations, which
could attract tiger attention. Such is the risk to honey
collectors from tigers, that their product has become
known as 'blood honey'.[21]

Fishermen are the least likely to be killed in terms of
the percentage of active fisherman that end up as
victims, but they account for the most attacks overall
because so many people enter the forest to fish. Given
that fishermen are working mostly from boats, and that
water might be considered to give people some
protection from attack even from a cat that is happy
swimming, this high number of attacks is surprising
and it certainly needs explanation. The nature of the
attacks actually allows us to find out some of the finer
details more easily than might be imagined. At the
moment of the attack only six victims (3.3 per cent of
the total in the study) were alone enough that no one
nearby witnessed the attack. The people using the forest
know well of the danger from tigers, and wood and
honey collectors move in groups of six to nine
and fishermen in two to three per boat. Since attacks

happen, this tells us that simply being in a group doesn't stop you being attacked, but equally 'being in a group' in the forest is not the same as being in a tightly knit group. The study concludes, on the basis of the data collected and insights from focus groups, that when a tiger selects someone it will follow that person until an opportunity to attack presents itself, more or less regardless of whether that person is notionally in a group, something that Mathew has also said to me. If people stick together very rigidly then attacks are unlikely to occur, but that is not the reality of collecting resources in the forest.

Having witnesses available allowed deeper analysis into what people were doing just before they were attacked. And those insights tell us a great deal about why people become vulnerable even in groups. When people's pre-attack activities are analysed it becomes clear that these are far more important than the reason why they are in the forest in the first place. Most people (76 per cent), regardless of what might be considered their occupation or reason to enter the forest, were attacked while collecting firewood or palms, or fishing; but close to half of those attacked while fishing were not fishing from a boat. Instead, these fishermen were explicitly described by witnesses as pulling in nets along the bank. It isn't hard to imagine why this activity would leave you vulnerable to attack. The researchers believe this is likely to be the case for most of the fishermen who were attacked.

With such a large dataset it, sadly, becomes possible to look at more nuanced aspects of the human element of attacks, including age demographics of victims. Echoing an earlier point, tigers are clearly selective predators and choose their victims carefully. Human beings are not all equal to a tiger and people between the ages of 15–29 are the least vulnerable. Tigers select them as victims far less than would be expected given how many people in that age range are in the forest. On the other hand, people aged over 45 and especially those aged over 60 are preferentially attacked as prey. Tigers are selectively hunting older people, who they presumably view as easier prey.

The Sundarbans data also allow us to make some inferences about tigers and their behavior. Overall, the findings there tie in with the pattern outlined by Mathew for other parts of India. Across a 23-year period in the Sundarbans it was estimated that 110 tigers killed people, averaging five victims per tiger.[19] However, and as is usually the case, the average (in this case the mean – adding all the victims up and dividing by 110 tigers) doesn't give us the nuance we need. More in-depth analysis reveals that around half of these tigers only killed one person, while tigers that killed more than one person accounted for 81 per cent of all deaths. The average duration of human-eating behaviour for these tigers is just over eight months, supporting Mathew's ideas that modern-day tigers grow out of the behavior in time.

As we saw in Chapter Two, people's attitudes towards dangerous animals – and the perceived risk they pose – can be at odds with the reality of risk. When the risk is real though, and the outcome fatal, it seems reasonable to expect some risk inflation. In the Banglasdesh Sundarbans, despite the high number of attacks, the reality is that most tigers aren't hunting and eating people, and that a great many people enter the forests without ever even seeing a tiger. However, of the 180 people surveyed in one study, 128 (71 per cent) believed that all tigers are 'man-eaters'. The answers to questions put to local people in focus groups show clearly that people understand that they are 'meat' and that tigers are predators. There is a sense of resignation, perhaps, that comes from living a life that requires you to enter a forest where tigers live, and regularly kill and eat people. The words of Hadas Kushnir and Craig Packer, summarising their findings on perception of risk of lions in Tanzania, ring true here too: 'People may not be responding to actual objective risk of death or injury, but to a deep generalised fear of predatory species.'

The misery inflicted on the people of the Sundarbans as a consequence of tiger attacks extends far beyond what we might expect to be the fallout of such an event. Obviously, there is the immediate loss of life and the trauma experienced by those who are present during the attack. Tigers carry off victims and eat part or most of them, and to recover the remains requires tracking the tiger (at great risk) and the drag marks made by the body. The success of this operation is measured by finding the horrifically mutilated remains of a fellow human being, quite possibly a family member or friend. The majority of attack victims in this area are men and in many cases these men will have families that rely on them economically. Tigers preferentially kill older people and when they do communities may lose valuable experience, wisdom and leadership. Within these communities – which along with those living with lions in parts of Tanzania experience the greatest predation by big cats of any communities globally – there is also an understandable sense of fear. Living in fear can be life limiting, and living daily with the stress of imminent attack (perceived or real) carries implications for physical and mental health. There is, though, a further cost in the Sundarbans that can potentially prolong the misery of a single tiger attack for many years; the creation of tiger widows.

The Sundarbans is a socially conservative region, traditional and far from the influence of modern-day India's urban middle classes. People there live much as they have always lived, and in these communities there is a stigma associated with being a widow. Traditional, patriarchal societies impose constraints on widows, who are denied access to many parts of social life because they are deemed to bring bad fortune. They are ostracised, marginalised and, in some cases, assaulted and humiliated. This is not an uncommon pattern across the world, and widowhood can carry with it immense physical, psychological, societal and economic burdens. In the Sundarbans these burdens are increased if the husband was killed by a tiger because of the additional religious significance associated with such a tragedy.

The deity Banbibi is the deity of the Sundarbans. Banbibi, the lady of the forests, is venerated by Hindus and Muslims alike in the region and is called upon by those entering the forest to protect them against tigers. The demon king Dakkhin Rai (the Lord of the South) is Banbibi's arch-enemy and is believed to take the form of a tiger that attacks people. The cult that has grown up around Banbini is incredibly complex, embracing spirituality and the ethics and use of the forest environment. To have faith in Banbibi is to validate ethical virtues and, over centuries, the veneration of Banbibi has become a forest religion. People venerate her for safety in their collection of forest resources, and within the myth, culture, spirituality and religion of Banbibi, tigers are symbolised very much as protagonists. Thus, tiger attacks are viewed as an indication of Banbibi's infuriation with the victim, and by extension the victim's family. Or, the widow can be actively blamed for the attack. Within the Sundarbans then, the already perilous condition of widowhood carries further religious and social costs when a tiger is involved in the husband's death. The tiger widows of the Sundarbans suffer throughout their lives from the stigma, which includes increased risk of suicide. Tiger widows often also face sexual assaults by their in-laws, who accuse them of being a *swami-khego*, or 'husband-eater'.[22]

What can be done?
It is tempting to make sweeping generalisations about tigers and people, but to do so ignores the considerable variation in geography and tiger ecology that we can find across areas in which tigers occur, and the differences that exist in the customs and practices of different human populations. The struggles of the people of the Sundarbans have parallels with those living around Bardiya National Park, and many of the root causes of human–tiger conflict are shared but, as the existence of tiger widows reveals, the details often differ. The same seems to be true when we consider tiger-related factors. For example, in Chitwan National

Park in Nepal, a study of 28 years' worth of data on tiger attacks suggests that, in Chitwan at least, the Corbett 'injury hypothesis' has some merit. In this region, 36 tigers killed 88 people between 1979 and 2006, and 56 per cent of tigers that were examined after being shot showed physical deformities, including damaged or missing teeth and gunshot wounds.[23] We can also see similarities with the situation in Bardiya National Park, with both areas suffering an increase in tiger attacks in more recent years. In Chitwan, this increase was pronounced, going from 1.2 people a year killed before 1998 to 7.2 after 1998. The reason for this rise? A tenfold increase in attacks on people collecting grass and fodder in the buffer zone around the forest. One theme that seems to run throughout human–tiger conflict is that of successful forest restoration or habitat protection leading to increased tiger population, coupled with the need of people living around or within protected areas to extract natural resources. Even when the details differ, as we saw with lions in Tanzania and other parts of Africa, it is inevitably the marginalised rural poor that bear the brunt of globally supported predator conservation measures.

As we have seen with lions, keeping people and livestock safe can be achieved, but it can be difficult and is seldom completely effective. For people venturing into the forests of the Sundarbans, Bardiya National Park or other prime tiger habitat to collect wood, palms or honey, or to fish, it is very difficult to prevent tiger attacks through simple, practical interventions. Measures that individuals can take against a powerful predator extremely well adapted to hunt with stealth in a complex environment like a forest are few and far between. Larger and more coherent groups could reduce risk and keeping older people out of the forest may also help, but in practice neither of these measures is likely to be practical. Human social behaviour might be some defence against predators, but our ability to make and use tools is also important. The best defence against a predator like a tiger is a firearm, but this is not an option for poor people working in forests. In any case, without training and practice the likelihood of someone using a rifle effectively in a sudden tiger attack is virtually zero. Corbett was a skilled and experienced hunter and soldier using well-maintained and expensive rifles, and he missed tigers in the heat of real-world encounters. Likewise, people may be carrying machetes or other tools that seem like a chance to even the odds in a tiger attack, but you still wouldn't bet on the human being. I wouldn't even bet on someone having the time or wherewithal to take a swing at a tiger in the second or so between realising it was coming for them and it being too late.

Overall, practical individual interventions are unlikely to be much more effective than venerating Banbibi before entering the forest. More effective measures have been proposed that focus instead on tigers and managing their interactions with humans. One suggestion is to impose a system of zoning in an area so that people can be excluded from zones where human-killing tigers are active. Regular patrolling and the sort of hazing tactics involving firecrackers and flares that we saw with lions could be used to discourage tigers from areas where people are present. Such a system could work, but would rely on good knowledge of tiger location and behaviour, as well as considerable effort on the ground. It also requires compliance from people, which is a major flaw in many a plan.

Another suggestion is to collar human-eating tigers with a tracking device and to monitor their movements. In principle this is a great idea. The location of the tiger would be known in real time and people could avoid areas where it was active. By involving local communities with the monitoring, such a scheme could build engagement, understanding and a sense of stewardship. Despite some potential, the difficulties with such an approach are many. First you must identify the problem animal and capture it, both steps of which are every bit as tricky as you might imagine. Because 50 per cent of human-eating tigers only killed one person in this area, but the other 50 per cent of tigers accounted for 81 per cent of victims, then really you should wait until a tiger has killed at least two people before undertaking the huge effort of collaring and monitoring it. Now, I don't know about you, but if a tiger had killed one of my children I am not so sure I'd want to wait until it killed another before anything happened. And, if it did kill again, and this known human-killer was on the floor sedated, I am not sure I would want to trust a collar-and-monitor scheme to save further loss of human life. I think I'd want to remove the problem while it was lying sedated on the floor. Can you honestly say that you would feel differently?

Collaring schemes are trying to achieve two things at once; protecting people and preventing the loss of a tiger. But what message does this give out to local people? Consider the duties of the Tiger Response Team (TRT) that was set up in the area by the Bangladesh Forest Department and two external NGOs, the Zoological Society of London and WildTeam, a UK charity focused on tiger conservation in the Sundarbans. These duties are listed as, '(1) Providing first aid to tiger attack victims; (2) Providing tiger attack victims with rapid transport to medical facilities; (3) Tiger victim body retrieval and transport; (4) Scaring tigers away that have strayed into villages.'[24] The final duty could prevent an attack, but the bulk of the TRT's work is about clearing up the mess afterwards. Tiger attacks commonly result in very rapid loss of life, or injuries so severe that loss of life is inevitable without immediate specialised medical help; first aid or rapid transport to medical facilities may well save a few lives, but it is closing the cage door after the tiger has bolted and caused massive injuries on the way out. Retrieval of bodies is a horrible task and by undertaking it the TRT is undoubtedly providing a valuable service, but once again, it is after the fact.

The problems of balancing people and wildlife

I believe we should be firmly focused on preventing attacks and protecting people, but navigating modern-day conservation and the attitudes and ideologies that underpin it is far from straightforward. Broadly, we can recognise two factions operating under the banner of 'conservation'. These factions arise from very different philosophies, ideologies and attitudes to the natural world. These differences often result in open conflict and yet it would be possible to be very interested in conservation without realising that this conflict lurks just beneath the surface. Because it is important to understand these factions in order to understand predator conservation – and because tigers are a good case study – now feels like the best time to tackle this issue, head on and openly. So, here goes...

Crudely, people and organisations interested in conservation tend to be either 'human-habitat-wildlife' focused or 'individual animals/animal rights' focused. There is of course a great deal of nuance in each position; both have their place and it is possible to have a foot in each camp. However, in general (and I stress, in general) those taking an animal-rights perspective put the rights of animals to exist at least equal to the rights of people. Organisations adopting this approach will often rally behind images of well-known and charismatic large mammals; I have yet to see anyone argue a rights-based approach to the conservation of spiders or mosquitoes. Animals are often identified as individuals, and may well be named. The rights of any single animal may be seen as equal to the rights of multiple animals, including whole populations, and of people. Those adopting this viewpoint will often consider their perspective to be intrinsically more moral than other approaches. Thus, many of those adopting this viewpoint would consider that killing feral cats on an island to save nesting birds is not just an undesirable action, but would be morally wrong. Humans are almost universally seen as 'the problem' (hard to argue with this), and the solution is often human exclusion through protected area approaches and complete non-utilisation of animals (*i.e.* no hunting for meat, no use for skins, *etc.*). The alternative view focuses more on protecting habitats and enhancing human–wildlife coexistence, with the 'rights' of ecosystems, habitats, species and communities dominating over the rights of any individual animals. This viewpoint often (but not always) accepts utilisation of animals for meat and other products. Those in this camp tend to view conservation as 'habitat centred' rather than 'animal centred', although in reality much attention is still paid to – and effort devoted to – the conservation of animal species.

I stand firmly in the latter camp. I do not think that any animal life is worth the same as a human life. Neither do I think it is de facto morally wrong to engage in activities such as the lethal control of invasive species. In fact, by ignoring human rights, prioritising individual animals over habitat and promulgating a view of the natural world that is at odds with reality, I think that the animal-rights agenda has been, and continues to be, harmful to real-world conservation efforts. I am very far from being alone in this opinion. I have yet to have a conversation with a single conservation scientist or practitioner who is not wholly or mostly in the same camp. I need to recognise, though, that my sample is biased by the fact that I tend to avoid conversations with those shouting loudly under a banner of animal rights, usually accompanied with a conveniently placed 'donate now' button. However, much global conservation focus is directed by those who shout the loudest, and many NGOs that purport to undertake conservation are underpinned by an animal-rights agenda. Such groups may also have considerable influence politically and enjoy plenty of media attention.

I have a deep-seated love of the natural world; that has always been part of who I am. I abhor animal cruelty, whether directed at a wild animal or at a pet. I do not cry for the death of a wild lion (as many claimed to over Cecil), but the loss of habitat in parts of the world I know and love does keep me up at night. I also believe it is unacceptable to expect people to live in constant fear and real risk of predation. For these reasons, I would advocate that if a tiger (or other predator) has a known record of eating humans then it should be shot, or that at the very least this should be a serious consideration. My position is not one rooted in a misguided sense of revenge and neither is it wholly motivated by preventing future attacks. My position, and that of many others, is that killing a human-eater swiftly can provide conservation benefits that far exceed the benefit of keeping a dangerous predator alive.

The logic behind this seemingly perverse position is simple: killing a human-eater now prevents many other retaliatory deaths later. People will turn against tigers if they are losing community members to them and once a mob has formed any tiger will do. Angry mobs armed with sticks can easily corner and kill a tiger, and they do. Meanwhile, the actual human-eater kills again, leading to more tiger deaths. The same thing happens with lions in Tanzania, which is why Amy Dickman supports killing human-eaters swiftly and openly, so that everyone knows the job has been done.[25] This approach was being taken in parts of India where the law gives chief wildlife wardens and senior federal officials the authority to issue orders to shoot tigers if it is warranted in the interest of public safety. This is not a small thing, but as Anup Kuma Nayak, additional director general of India's National Tiger Conservation Authority, told journalist Rachel Nuwer for the BBC in 2019, 'If a tiger is really a man-eater, we have to go after this man-eating tiger according to a very well-defined standard operating procedure.'[26] But in India, animal-rights groups are protesting this approach. Based in urban areas, with influence and resources, their voices are far louder than those of people living with tigers. These people want nothing done, or for animals to be trapped and relocated, or kept in captivity (with little regard for what such an approach would entail). The result has become a political 'hot mess' and no permits have been issued since 2018. Protracted legal battles and pushes for 'humane' approaches have come to dominate the tiger narrative in recent years, with the inevitable result that local people are sometimes taking matters into their own hands.

Tigers are a conservation success story and long may this continue to be true. But for me, this success transcends simple population increases. They are a success story because population increases show us the way forward for conservation. Communities are the key, and when communities are empowered and incentivised to protect habitats and wildlife they can do so in ways that defy even the harshest critic. However, if we accept that communities are the key to conservation then we must also accept that communities have the right to govern and manage their wildlife populations in ways that are right for them. It is wonderful that people in London, Paris and New York want to save tigers, but converting that desire to meaningful action on the ground requires far more engagement with the complexities of real-world conservation than simply clicking a donate button. Of course, not everyone can actively participate in conservation, and funding is vital, but if you want to save species and habitats make sure that your well-meaning donations aren't inadvertently tying the hands of the very people you are trying to help. Support initiatives that work with, not against communities, and those that respect, not diminish, human rights. Species, habitats and people will thank you for it in the end.

CHAPTER FOUR

Crocodilians

One of my earliest memories is one of fear. I was somewhere between three and four years old, visiting Paignton Zoo in Devon. It was a warm day and I remember being pushed around the seemingly endless concrete paths in a pushchair. I think I can remember constantly asking to see the crocodiles, but that may be an invented memory seeded by later conversations. What I definitely do remember is finally seeing a crocodile in the tropical house and bursting into tears. That huge, unmoving and unnerving beast lying on the ground next to a pool of water was, without doubt, the most terrifying thing I'd ever seen. Some 40 years later, I took my children to the same zoo and, in a very much refurbished tropical house, my four-year-old daughter had exactly the same experience. Given that several species of crocodile can live in excess of 50 years it may even have been the same animal.

Why did I have such a visceral and memorable reaction when I first saw a crocodile up close? Perhaps the key to answering that question has to do with the fact that there is something very different about a crocodile than, for example, a big cat. When we look at a lion or a tiger we are looking at a warm-blooded mammal with which, in many ways, we can identify. Lions are social animals with interactions we can interpret through familiar lenses. Sure, our interpretations can be romanticised and wildly incorrect at times, but when we see lion cubs playing with an adult lion our anthropomorphic analysis of the situation may well be more or less on the money. When we imagine a tiger roaming great swathes of forest, silent, alone and majestic, we project our own aspirations and feelings on to it. When we look into the eyes of a big cat we can sense that 'something' is behind them, even if we are not sure what that something might be, or how to express it. The same cannot be said for a crocodile. A crocodile is not a warm, furry animal, with a life that has aspects with which we readily identify. It looks like exactly what it is: a predator. And, to a crocodile, you look like meat.

Crocodilian diversity

The Nile crocodile I saw at the zoo is just one of a group of species of semiaquatic reptiles collectively known as the crocodilians and found throughout the tropical regions of the Americas, Africa, Asia and Australasia. Three species in particular will be the focus of this chapter, the Nile crocodile of Africa, the mugger crocodile of India and the saltwater crocodile of Australasia, but the other crocodilians are worth a quick detour. Not least because several of them are known to eat us.

The extant (as opposed to extinct) crocodilians are split into four main groups. The crocodiles are 17 species that include the three mentioned already, as well as species with common names mostly reflecting where they are found: the Borneo crocodile, the Philippine crocodile, the Cuban crocodile, and so on.

Superficially similar to crocodiles, but generally smaller, the crocodilians also include the caimans (or caymans), with a handful of species found from Mexico throughout Central America to northern South America. Although it is relatively small, the black caiman can reach 4m (13ft) and 500kg (1,102lb), which is a big animal by anyone's measure. On the other hand, the smallest species, Cuvier's dwarf caiman, doesn't get longer than 1.4m (4ft 7in) and tips the scales at just 7kg (15lb). A dwarf caiman can of course give you a nasty defensive bite, but a 7kg 1m-long reptile isn't looking at you as prey. The same cannot always be said of the black caiman. Although attacks are considered uncommon, even in areas where caimans and humans are present in high densities, there are well-documented examples confirming that predatory attacks occur.[1]

In February 2010, seven children were playing on the side of the river in Guajará-Mirim, a town in Rondônia State in the Brazilian Amazon. Around midday one of the children, an 11-year-old girl, was attacked by a black caiman, which took her in its jaws and disappeared. Later that evening, around 100m (328ft) from the attack site, a 4.2m (13ft 9in) 350kg (772lb) black caiman was located. Surfacing with the girl still in its mouth, the animal was shot and killed. An autopsy showed that the caiman had grabbed the girl by the thighs and dragged her under, where she drowned. This is the typical profile of a crocodilian attack on larger prey. The seasonal flooding that occurs throughout the area is now identified as a risk for caiman attacks, which thankfully remain rare.

Switching to a better-known crocodilian, we come to the alligators. Most people are familiar with the American alligator – the 'gator – but there are actually two species, the second one being the Chinese alligator. In what is now a familiar story, the Chinese alligator was once relatively common and widespread, but through persecution, hunting for meat and traditional medicine, and habitat destruction Chinese alligators now number in low triple digits in the wild. Listed as Critically Endangered by the IUCN, the Chinese alligator is too small (around 1.5m/4ft 11in and 40kg/88lb) and too rare in the wild to worry us in terms of human predation. Lest we get too down in the dumps about its prospects, though, there are at least 20,000 in captive-breeding centres and zoos around the world, and reintroduction programmes are a high priority. It is a sad reflection on the human condition that we can create such successful captive-breeding programmes while simultaneously continuing to do the very things that make such programmes necessary.[2]

The American alligator, in contrast, is listed as being of Least Concern and has an increasing population, doing well through the US states bordering the Gulf of Mexico and northwards into the Carolinas.[3] This was not always the case and in the 1970s alligators were listed on the US Endangered Species Act. Subsequent conservation efforts worked well (being large, culturally important and well known never hurts when it comes to conservation) and now, thankfully, alligators are off the sick list. However, being of Least Concern from a conservation perspective does not mean that the

American alligator doesn't have its threats and, no surprises, habitat destruction looms large. But for now, it is an example of how conservation efforts can reverse worrying declines.

American alligators are around the same size as – or slightly larger than – black caimans, with bigger males comfortably exceeding 4m (13ft 1in) and 400kg (882lb). Alligators often live in areas of relatively high human population density, so people and large alligators are likely to come into contact with one another. Despite this overlap, alligator attacks and fatalities are uncommon. A 2010 report collated all known attacks from 1928 to 2009 and found 567 reports of 'adverse encounters' with alligators and 24 deaths, mostly (22 of the 24) in Florida.[4] This works out as one death every three years. Adverse encounters included bites and a later analysis in 2019 focused only on bites in Florida, where most occurred.[5] This identified 372 bites between 1948 and 2014, which works out to just over 5.5 bites a year. Total deaths may be higher than 24 because it is possible that some victims weren't recorded, or listed simply as a missing person. However, the US has better record-keeping than many places in the world, and it seems unlikely that this number is as problematic as the estimates we met in earlier chapters. Each death is a human tragedy for the victim and the families, of course, but given the number of alligators and their likely rate of interaction with humans, a death every three years is surprisingly low, even given the period when alligators were on the endangered species list.

It is worth dwelling a little on the American alligator. This is an animal that is definitely capable of killing and eating humans, and is common in areas where people live. Speaking to the BBC, Lucas Nell of the University of Georgia said 'it is reasonable to assume that *any body of water in Florida* could have an alligator in it' (my emphasis).[6] Yet despite motive and opportunity, the risk from alligators does not translate to the levels of harm we might predict. We will shortly see the problems that its relatives, especially the Nile crocodile, cause in similar high-opportunity scenarios, but the crucial difference between crocodiles and alligators is that alligators are picky eaters. Crocs are dustbins and will eat anything that is within their prey spectrum, including larger animals like us. Alligators, on the other hand, target smaller animals, with a diet comprising fish, other reptiles like freshwater turtles and terrapins, birds and small mammals. When alligators do attack, only 6 per cent of attacks are fatal, compared with 63 per cent for Nile crocodiles.[7] Most alligator bites in Florida do seem to be related to feeding (according to the 2019 analysis of bites) but more than a third (37 per cent) of incidents probably weren't, taking the form of a single bite and immediate release. This 'bite-release' behaviour indicates second thoughts or a defensive bite. In general, it seems that we mostly don't look like convincing prey to most alligators, which is fortunate. Top safety tip, though: don't swim in bodies of fresh water in Florida; and alligators, like most animals, are best viewed from a distance. Don't try to get closer to them and, although this shouldn't need to be said, don't try to grab them or otherwise interact with them. They don't like it and it serves no purpose. Poor practice on the TV doesn't need to translate to tragedy in real life.

Finally, in our line-up of extant crocodilians, we come to the weird and wonderful gharials. The species we've met already all look (and apologies to crocodilian anatomists) pretty much the same. They are mostly distinguished by important differences that can be difficult to see, including teeth placement and the presence or absence of sensory organs around the jaws. The most obvious difference between them is in jaw shape, with crocodiles having a more pointed jaw than the broader-jawed alligators and caimans, but even this can be tricky to judge from certain angles. However, when it comes to distinguishing different crocodilians from each other on the basis of jaw shape, the gharials pose no problem at all – even to the untrained observer and from pretty much any angle. Gharials, or gavials, have distinctively long and very slender snouts with a bulbous 'pot' at the end, especially prominent in sexually mature males. It is this feature – and its resemblance to the earthenware pot of India and Pakistan known as a *ghara* – that gives the gharials their name.

There are two species in the gharial group, the gharial and the false gharial. The gharial is found mainly in India (with small numbers in Nepal), while the false gharial is found in Malaysia, Sumatra, Borneo and Java. The gharial is Critically Endangered, reduced to only 250 individuals by 2006 through persecution, collection of eggs, and hunting for meat and medicine. It has also been eradicated from all but a tiny fraction of its former

range. It is all so depressingly familiar, isn't it? False
gharials are doing a little better, and there is hope from
some captive-breeding and reintroduction programmes
for gharials in India and Nepal. However – and prepare
yourself for yet another familiar trope in global
conservation – attempts at reintroduction have often
failed in large part because the efforts to reintroduce
gharials have become institutionalised rather than
engaging with local communities.[7] Remember that
Nepal tiger model? We really do need to learn from that.

Gharials are the most aquatic of all the crocodilians
and their slender jaws are well adapted to catch fish.
Despite weighing 150kg (330lb) and reaching lengths in
excess of 3m (9ft 10in), the lifestyle and feeding ecology
of the gharial means that humans are not really on its
prey spectrum and gharials are not generally considered
to be dangerous. That said, a 3m reptile with a front end
loaded with teeth is more than capable of inflicting a
nasty defensive wound, and such a wound could quickly
prove fatal if a major blood vessel was hit, or it could lead
to a slower death should it become infected. This applies
to any of the crocodilians, but such attacks are not
predatory in nature. The false gharial, on the other hand,
is a larger animal, reaching 5m (16ft 5in) and 200kg
(441lb), and there are credible reports of it eating people.

The first verified report of a predatory attack occurred
in November 2008 in Pangkut, in the Kotawaringin
Barat district of the Central Kalimantan province in the
Indonesian portion of the island of Borneo. An oil-palm
plantation worker swam across a river taking a shortcut
home and was taken by a crocodilian, most likely a false
gharial. A week-long search was initiated that failed to

find the animal or any remains, at which point a local *pawang* (shaman) was called to help. As luck would have it, this pawang used to be a professional crocodile hunter and was successful in finding and killing a large (more than 4m/13ft 1 in) male false gharial. He was convinced that this was the killer, but a dissection of its stomach did not reveal any human remains. Incredibly, on New Year's Day that year another attack happened. A 43-year-old man was pushing a raft of logs along a river when he screamed and disappeared. Two companions were unable to find the man. Returning with some 100 others, human body parts were located but no crocodilian. The next day, the pawang who had followed up the earlier attack joined in the search and, maintaining his 100 per cent kill record, a 4m-plus female false gharial was stunned with electricity and dispatched with spears. Subsequent dissection revealed three primates in the gharial's digestive system; a proboscis monkey, a long-tailed macaque and the missing man.[8]

No further gharials or other species were killed following this attack and this is worth noting, especially with regard to some of the themes I started to explore in Chapters Two and Three. It is entirely natural to wish to find the animal that has killed someone close to you, or a community member, and it is entirely natural to want that animal dead. Killing the animal satisfies a number of human needs, including the need to be completely sure what happened (the need for closure), the need to have some mortal remains for an after-death ceremony, the need to prevent further attacks and, bluntly, the entirely understandable need for revenge. There is another report of a crocodilian attack in Sarawak (the Malaysian part of the island of Borneo) that triggered widespread and indiscriminate killing of all crocodilians, which led to considerable conservation concerns. With the New Year's Day attack, being able to track down and kill the animal quickly and to demonstrate beyond any doubt, as grisly as that demonstration might be, that the right animal had been killed did much to defuse a situation and prevent further animal killings and animal suffering. The advantages of rapid and definitive action were also expressed to me by Amy Dickman regarding the aftermath of lion attacks in Tanzania,[9] and by Rajeev Mathew when speaking about tigers in India in Chapter Three. As a general principle, it seems applicable to any predatory killing of a human by an animal. Acting quickly, and killing the right animal, can undoubtedly save many more animals from ill-targeted retaliatory killings and may stop future deaths of humans. To put it bluntly, if you elevate the rights of that single animal to live above all else, you may well be responsible for multiple animal and human deaths. Standing on the sidelines and criticising local people for killing a predator that has taken one of their own (as I have seen many times on social media) shows a lack of empathy for sure. It also reveals a lack of thought about a far more complex picture than a photo of a dead animal and some grim-faced people carrying sticks.

Crocodile attacks

On 8 June 2021, I opened up the BBC news website to see, nearing the top of the most-read list, a particularly arresting headline: 'Brit saves twin sister by punching crocodile in the face'.[10] To be honest, the headline does much of the heavy lifting here, but the details of the attack were quick to come in a story that was picked up and told all over the world's media. Twenty-eight-year-old Melissa Laurie, from Berkshire, UK, was swimming near Puerto Escondido on the Pacific coast of Oaxaca state in Mexico when a crocodile dragged her underwater. Her twin sister Georgia punched the animal in the face, more than once, and eventually Melissa was saved after nearly drowning. Melissa was reportedly put into an induced coma because of concerns over infection, but recovered.

I know the area where Melissa was attacked. It is a nice spot and the mangrove lagoon, Laguna Manialtepec, has become something of a tourist destination in recent years. It was a bit more laid back when I was there in my twenties. I spent the best part of six months in Mexico and Central America after I'd finished my undergraduate degree and, a few years later, I acted as a tour guide for beekeepers exploring Mexico. I saw a fair few crocodiles during this time and I also saw a fair few people behaving without due regard for the danger that crocodiles pose. In one case, a German tourist keen to get a photograph (in the days before mobile phone selfies) approached a large American crocodile, the same species that attacked Melissa. His wife urged him to get closer. Coming to within a couple of metres of the animal's head, he crouched down with a massive smile on his face. That smile disappeared very quickly when the crocodile, which hadn't moved a millimetre for at least an hour, arced its head very swiftly, jaws opening, towards the man. He was lucky – it was a warning, but really such a warning should not be necessary.

There are three stark differences between the Melissa Laurie attack and most other crocodile attacks. For a start, this was a 'recreational attack'. The victim didn't 'need' to be in or near the water. The Laurie twins weren't fishing, tending nets, collecting water or engaging in the many other mundane tasks that routinely force many who are attacked to be near water bodies with crocodilians. Secondly, she was lucky to be able to be treated at a good hospital. Thirdly, just like the story of Gotz Neef who was attacked by a lion in Botswana (see Chapter Two), this story became big news globally. 'Westerner survives wild animal attack' is clickbait. An additional difference is that the species that attacked her, the American crocodile, is not the species responsible for most attacks. That honour falls to the Nile crocodile.

Nile crocodiles are found in sub-Saharan Africa and are relatively common. Unlike the American crocodile and the saltwater crocodile, both of which are frequently found in the sea, Nile crocodiles are a freshwater species. They are, however, pretty robust, tolerant of all kinds of conditions and have been found swimming in the sea at times.[11] The fact that they can tolerate a broad range of conditions, including estuaries, swamps, fast-flowing rivers, lakes and dams, means that Nile crocodiles are very widely distributed, in terms of both geographical area and habitat. If you are in sub-Saharan Africa and there is water around, it is a fair prediction that they could be present. You would certainly be best to proceed under that assumption before swimming.

As well as being widely distributed and relatively common, Nile crocodiles are large. Measuring crocodiles reliably, especially if they aren't dead, is tricky. That, coupled with our fascination for record-breaking animals and a tendency to exaggerate, leads to some confusion over the size of the Nile crocodile. 'Typical lengths' are reported as anywhere between 2.8m (9ft 2in) and 5.5m (18ft). One study of the crocodile population of Lake Turkana in northern Kenya in the 1960s concluded that 'males seldom exceed 470cm [15ft 5in] length and females 320cm [10ft 6in]', with an overall average of 3.66m (12ft) and a weight of just over 200kg (441lb).[12] This feels like a reasonably secure estimate of size, but there are credible reports of Nile crocodiles exceeding 6m (19ft 8in) and 1,000kg (2,205lb). That brings us nicely to Gustave.

Gustave is (or possibly was...) a Nile crocodile from Burundi that has been estimated to be in excess of 5.5m (18ft) long and likely to weigh more than 900kg (1,984lb). It takes a while to grow that big and Gustave may well be more than 60 years old. His length, weight and age are pretty impressive, as are the legends that surround his human-eating exploits around the Ruzizi River and the northern shores of Lake Tanganyika. He is rumoured to have killed more than 300 people, including, apparently, the wife of the Russian ambassador to Burundi.[13] He has been active since at least 1998, when the first reliable reports came in of a crocodile targeting fisherman freediving to collect cichlids for the pet trade. However, there were even earlier reports, dating back to 1987, of a huge crocodile attacking and killing people. Through a process of elimination and deduction based on Gustave's presence and absence in different areas, he was implicated in those attacks too. He may have been killed in 2019, but reports of his demise have, to date, not been confirmed.

The predatory behaviour of Gustave the Nile crocodile repeats a pattern we have already established with lions and tigers. The victims are the rural poor, they are going about their daily lives, the precise number of victims is unknown and in most press write-ups of 'Gustave the killer croc' the victims are unnamed. Gustave seems to be unusual in his predatory behaviour, and the presence of three bullet wounds on him have led some to speculate that injuries could have led to a focus on human killing as an easy option; an echo of the Corbett injury hypothesis we met in Chapter Three. However, this is not likely to be the case for most Nile crocodiles that attack humans, or indeed most other crocodilian attacks.

Just how many people are attacked and killed every year by Nile crocodiles in Africa is hard to pin down with any certainty, but the number is certainly in the high hundreds and may be much higher. Overall, we can be confident that the Nile crocodile kills more people than any other predator in Africa, but this statement doesn't really tell us much about where attacks happen, and why. To drill down into the detail of attacks requires data and these are sadly lacking. Of the 30 African countries where bites are known to occur, we have data for just 12 of them. Most of the information we have covers short periods of time and was collected in different ways.[14] It is tempting to think of attacks spread out uniformly across the landscape, the primary factors in their distribution through space and time being the opportunity for crocodiles to interact with humans. However, crocodiles are not primed to attack any human that comes within striking distance and many areas with crocodiles may experience no attacks at all. On the other hand, some areas may have a long history of adverse encounters. For example, the Zambezi River was recorded as being dangerous because of crocodiles by David Livingstone in the 1850s.[15] It has become a hotspot for attacks again, with multiple reports emerging of attacks over recent years, many of which were fatal.[16]

Without good, long-term data across their range we are limited to our attempts to understand crocodile attacks and to put them into a wider context. Crocodile ecology and behaviour don't help us much here either. Lions and tigers have a relative rich behavioural repertoire and aspects of their lives, such as group living (or not), territoriality and dispersal give us something to chew on. Crocodiles, on the other hand, seem to do far less, or at least are far less studied. Overall, we have fewer leads as to why and when crocodiles attack. Linked with the reputation as 'cold-blooded killers' (a cliché that appears in many media stories covering their attacks) it is easy to fall into the assumption that they'll come for us if we give them half a chance. But we should be wary of simplistic assumptions.

One person who has tried to address the data shortage on crocodile attacks, and use an evidence-based approach to provide insights into how we might prevent incidents, is Simon Pooley of the IUCN Crocodile Specialist Group and the task force on Human–Wildlife Conflict & Coexistence. Together with colleagues, Pooley analysed 67 years of data on crocodile attacks from 1949 to 2016 in South Africa and eSwatini (formerly Swaziland). By searching through literature and various archives, the team identified records of 214 attacks that allowed it to examine patterns of where and when attacks occurred, and who was most likely to be a victim.[17] Nile crocodiles are capable of taking very large prey indeed, including cape buffalo, so any human being is well within its prey spectrum. However, in the South Africa and eSwatini data an interesting pattern emerged; the majority of victims (65 per cent) were male and most (51 per cent) were under the age of 15. Most victims (31 per cent) were attacked while swimming or bathing, followed by fishing (22 per cent) and domestic chores at the water's edge coming in a close third (18 per cent). Not unsurprisingly, children were significantly more likely to die from crocodile attacks than adults (54 per cent of attacks were fatal versus 35 per cent).

Analysing the data further, Pooley and the team were able to show that, despite crocodiles being relatively widespread in different freshwater habitats, most bites occurred in the warmer, low-lying eastern parts of the region, and in rivers that were linked to major crocodile populations in the St Lucia Lake system, Ndumo Game Reserve and Kruger National Park. They also found that bites were far more likely to occur in the summer months between December and March. Crocodiles may be more active then because of higher temperatures and, because it is the breeding season, more aggressive.

Stopping children swimming and bathing in certain rivers during the summer months could clearly reduce attacks. In practice, in dispersed rural communities, this may be difficult or impossible to achieve, but educational outreach directed at children and their parents, including advice on responding to and avoiding attacks, could be effective if it was carried out correctly. In specific high-risk areas safe water-crossing points could be established and water-delivery infrastructure, including water tanks and piped water, could be provided to reduce human presence on riverbanks. Protective enclosures that allow people to bathe and wash clothes safely could also be installed. As with educational outreach, though, this is all easier said than done; identifying effective mitigation does not make that mitigation a reality. Currently, policies are varied, inconsistently applied and reactive, focusing on problem animal removal rather than conflict reduction. All the time that debates over what should happen, who should pay and how it should work continue, so will attacks.

The pattern of attacks in South Africa and eSwatini is highly variable, with some districts seeing attacks decrease since 1949 while others have increased. Particularly notable is the increase observed in attacks occurring in dams. Dams are a frequent sight on properties throughout South Africa. Rivers are partially blocked off, typically with an earth-constructed wall, to create an upstream lake in which rain that falls (often in spectacular amounts) during the wet seasons can be stored for the drier months. It is the lake, rather than the wall, that is commonly referred to as 'the dam'. During the 67-year period between 1949 and 2016, 64 per cent of dam attacks happened in just the last 16 years, which is a dramatic change in the geography of attacks. There seem to be two key factors underpinning the increased attacks in dams. First, crocodiles are moving into dams because rivers are no longer good places for them to live. Perennial rivers are drying up, mainly because water is being diverted and removed for irrigation, and rivers in general are becoming more polluted or disturbed through human activity. This habitat shift, from rivers to dams, potentially brings crocodiles more into contact with people who use dams for all the usual things for which humans use large water bodies.

In South Africa and eSwatini we can see how the effects of human activity and infrastructure development, like farm irrigation and dams, result in changes in crocodile ecology that can lead to increased attacks on humans. But, regardless of why people and crocodiles end up in close enough proximity for attacks to happen, the key factor in crocodile attacks is always going to be people being close to, or in, water. Moving north and east into Mozambique, we can once again see the common pattern emerging with respect to who is likely to be in water. The rural poor bear the brunt of attacks there because it is the rural poor who have to use water sources that might contain crocodiles to find water, to fish, to wash clothes and to bathe. In a study of wildlife attacks in Mozambique in the 27-month period between July 2006 and September 2008, Kevin Dunham and co-authors evidenced 134 people (or around one person every six days) being killed by crocodiles and a further 36 people injured.[18] Attacks were concentrated in 12 districts lying alongside the Zambezi River and around the shore of Lake Cabora Bassa, but attacks were recorded in 46 districts across the length and breadth of the country, reflecting the widespread distribution of crocodiles and people. Interestingly, attacks on livestock were far less distributed across the country, and were only reported from districts in central and southern Mozambique. But what is more interesting still, is what happens when you look at the numbers of crocodiles killed in response to attacks.

In the same period during which 134 people were killed and 36 injured, 92 crocodiles were reported to have been killed as a consequence of human–crocodile conflict. If we just consider the human fatalities (rather than including injuries), this amounts to 'just over two-thirds of eye for an eye' and feels remarkably low given the horrific reality of a crocodile-inflicted death every six days or so. Regardless of the motivation for crocodile killing, the low numbers killed are worthy of discussion, especially when compared with other human–wildlife conflicts in the same areas. The same study publishes numbers of elephant-caused deaths and elephants killed; 31 human deaths, six people injured and 85 elephants killed in response. That is considerably more than 'two eyes for an eye'. The pattern continues with other species: 12 people killed by hippos, 60 hippos killed; one person killed by a buffalo, 11 buffalo killed. The reason that seems to underpin these retaliatory patterns tells us much about people's capacity to live with dangerous wildlife. The species that are the target of 'over-retaliation' (elephants, hippos and buffalo) are those that cause crop damage and livelihood losses. It is these losses and dangers, rather than the loss of or danger to human life, which seem to fuel retaliatory killings. This livelihood-damage idea is further strengthened by the fact that there is a weak correlation between loss of cattle to crocodiles and crocodile killing, but no correlation at all between the loss of human life and crocodile killing. If we can keep people from being so reliant on bodies of water for their basic needs, protect them when they are there and (with a knowing nod back to lion-proof *bomas*) protect livestock and crops, then we can reduce harm to people and subsequent harm to wildlife. Yet again, the inevitable conclusion is that increasing the economic prosperity and development of people who live alongside wildlife is the key to protecting them and the wildlife.

Why not kill them all?
When we consider crocodile predation, the loss of human life and the fact that people can and do kill them, it really is noteworthy that animals like crocodiles are still around at all and have not been completely wiped out. This would be perfectly achievable. After all, the decline of crocodiles in many areas attests to the ease with which we can eliminate them. Nonetheless, across its range the Nile crocodile is not uncommon and in some places exists in large numbers despite being a danger to people. I live close to the River Severn and the Avon in the UK, and the area around me is riddled with drainage ditches, streams, ponds and culverts. If crocodiles lived in the UK this would be as ideal a habitat for them as it is for the numerous dog walkers I see enjoying the views across to the Malvern Hills. How many crocodile-caused human deaths (or dog deaths for that matter) would we tolerate? I suspect the answer is none. So, why do crocodiles still exist in so many places

where people live? The simple answer is that the people who live alongside dangerous wildlife seem to be more tolerant of it than we, sitting in safe and comfortable houses thousands of miles away, might imagine. The reasons underpinning that tolerance are likely to be complex, tightly linked with local culture and history, and connected to the species itself. One area where an attempt has been made to understand tolerance in relation to crocodiles is the Korogwe district in Tanzania.

The Korogwe district is in the north-east part of Tanzania, just inland from the Indian Ocean coast. Richard and Heather Scott, publishing in the *British Medical Journal* (under the section heading 'Bites'), studied crocodile attacks that occurred in the 52-month period from the start of January 1990 to April 1994.[19] They collated records from the Korogwe Department of Natural Resources and concluded that there had been 51 human deaths during that period, starting with five in 1990 and rising to an alarming 18 in just the first four months of 1994. They did not include non-fatal injuries, the severity of which can be appreciated by looking at the photograph in the paper of a man with two huge, deep and gaping wounds on his back and down to his buttocks. His spine is just visible in places, surrounded by multiple smaller puncture wounds.

We can put the fatalities in 1994 particularly into a context that may be more meaningful. At the time of the study, the most recent Korogwe district census (taken in 1988) recorded 20 villages in the district and a population of 217,810. By 2012, despite a large increase in the overall population of Tanzania, this largely rural province had increased to just 242,038 people.

The current population of the UK is somewhere around 67–68 million, so there are about 280 times as many people in the UK as in the Korogwe district, give or a take a few. Imagine then that between 1 January and 30 April in any given year we had experienced 5,000 (280 x 18) deaths from crocodiles in the UK. How many crocodiles do you suppose we would have left?

The Scotts did far more in their study than simply collating death records. They also interviewed residents to try to get to the bottom of who was being attacked, why the deaths had increased so dramatically and, in their own words, why 'these voracious human preditors [sic] are not exterminated' because 'they seem to be flourishing at the expense of the local population'. The reason for the increase in deaths seems likely to be linked to river pollution and overfishing affecting the availability of fish and other prey, but the reasons why villages don't hunt crocodiles to local extinction are rather less prosaic. The Scotts conclude that traditional beliefs and superstitions linked to crocodiles are the key reason that they still exist in the area. These beliefs include the idea that some crocodiles are 'tamed' and that their owners are able to get the animals to do their bidding, including procuring women, assassinations and body disposal. Attacks are seen as being carried out by tamed crocodiles and the fear of 'witchcraft' (as the Scotts term it) used by their owners gives some degree of protection to the crocodiles. Other beliefs that act to protect crocodiles include the idea that it is impossible to kill a tamed crocodile and that if you kill a crocodile you will lose a son to another animal.

The Scotts concluded that these beliefs collectively allow a pervading fear to develop among the people of Korogwe that prevents individual and collective action to reduce crocodile numbers and attacks. They propose that the importance of education in 'countering the influence of superstition cannot be overemphasised'. By reducing fear, they suggest such education would 'allow crocodiles to be killed more freely', but deep-seated beliefs – however strange or ridiculous they may appear to outsiders – are rarely shaken easily. What is more, this aspect of the Scotts' solution is focused on reducing human–crocodile conflict by reducing crocodiles, which may solve the medical problem of bites but does not result in a good conservation outcome. Far more useful – and certainly far better for human–predator coexistence – is the practical advice the Scotts offer that aligns with the development theme we have already explored. The clue to reducing attacks without wiping out crocodiles can be found in the catastrophic fatality figures from early 1994. During this period a water pump failed and this meant that people had to go to the Pangani River to collect water. This simple technology failure escalated deaths from just under one a month in 1993 to about one a week. Water pumps save lives – and not just by providing clean drinking water.

Simon Pooley, the lead author of the South Africa and eSwatini study discussed previously, further explored the importance of beliefs in human–crocodile interactions across Africa in what he terms a 'cultural herpetology'.[20] In a fascinating study, Pooley highlights the importance of taking this broader approach in considering ways to mitigate human–wildlife conflicts, stating that 'conservationists wishing to intervene in human–predator interactions must be aware of the strata of personal, social, cultural and political meanings and relationships pre-existing their entry into local relations and practices'. Crocodiles have long been associated with strong spiritual beliefs throughout their range in Africa. Echoing some of the themes from the Scott account in Tanzania, a key conclusion is that, 'Death by crocodile attack is more often attributed to either justice being meted out to immoral humans, or evil humans bewitching or becoming crocodiles to murder or maim their enemies or those of their clients.' A strong theme that develops in Pooley's account is that, in general, Africans tend not to blame all crocodiles for attacks. Instead, in some societies there is a sense of a shared ethical framework, where crocodiles that kill humans are seen more as individual criminals to be punished. Some societies also consider crocodile attacks to be punishment meted out to a sinner or as a manifestation of ancestor anger, but again these explanations emphasise the action of a single crocodile, not all crocodiles, and place the blame (the reason for the attack) on the person. The crocodile is more of an instrument than an actor. It is easy to judge other people's beliefs, but what Pooley's work reveals is a complex system of beliefs across Africa that, overall, tend towards coexistence with these predators rather than extermination. Perhaps the rest of the world could learn a thing or two…

Salties

Before 1986, Australian actor Paul Hogan was known chiefly for his comedy sketch series *The Paul Hogan*

Show. A hit in his home country, the show was also popular the UK, where it was aired on Channel 4 as part of the fledgling channel's comedy offering. My dad was a fan and I can remember staying up late to watch Hogan as Sergeant Donger, a plain-clothes cop with a bionic beer belly. But, in general, Hogan was not especially well known internationally. All that changed in 1986 with the release of the movie *Crocodile Dundee*. Suddenly, Paul Hogan, and the saltwater crocodiles of Australia that his character Mick Dundee hunted (illegally), hit the big time. The fact that Dundee, a self-confessed crocodile poacher, is lauded as a hero in the film tells us a great deal about skewed conservation narratives. I cannot imagine a film making light of crocodile poaching in Tanzania, with a black 'hero poacher' making wise-crack remarks, being a blockbuster hit, no matter who played the lead role. Regardless, sequels followed and Hogan made a series of amusing and doubtless lucrative adverts for Foster's Lager. By the time Steve 'the Crocodile Hunter' Irwin came along with his TV series a decade later, the saltwater crocodile was firmly cast in the role of massive, aggressive and dangerous predator.

'Salties' are certainly massive. The largest living reptile, a fully grown male can exceed 6m (19ft 8in) and weigh 1,300kg (2,866lb). Thanks to film and TV, they are most associated in people's minds with the coastal regions of the northern portion of Australia, but in fact they have a widespread distribution through Indonesia and north to the Philippines, west up through South East Asia and along India's east coast as far as Sri Lanka. Although still widely distributed, as is the case with so many other species we are meeting, their current range is far patchier than it once was.[21] An apex predator, a mature saltwater crocodile can take down pretty much anything that comes its way, including other predators like the bull sharks that are commonly found in estuaries and further up river systems in Australia.[22] They aren't above a bit of scavenging either, with a recent report detailing saltwater crocodiles and tiger sharks feeding together on a whale carcass off the coast of Western Australia.[23] A large and powerful predator that coexists with humans across much of its range, like the Nile crocodile and the black caiman, salties clearly have the potential to make life hard for us. But just being big and having opportunity aren't enough to turn a predator into a problem. For that to happen, predators need to have motive, to view us as food. And salties most assuredly do. Professor Grahame Webb doesn't pull his punches when he sums up how they view us: 'There is no way of avoiding nor sugar-coating the predatory nature of saltwater crocodiles. If you dive off the Adelaide River bridge, 60km (37 miles) east of Darwin's city centre, and start swimming, there is 100 per cent chance of being taken by a saltwater crocodile. It is not the same as swimming with sharks.'[24] A 100 per cent chance. I don't know about you, but I don't like those odds.

Salties have a media presence that does nothing to allay fears. I'm writing this in July 2021 and the first news story I find about saltwater crocodiles is a gory tabloid piece published a few hours ago in UK newspaper *The Sun* detailing a series of attacks on people by saltwater crocodiles in Australia, Indonesia and Malaysia. With the usual breathy and hyperbolic tone of the paper, injuries and tragedy are described, and the animals responsible pronounced 'fierce', 'cold-blooded' and 'brutal'.[25] None of those words seems out of place, even with the gory parts pixelated. Scrolling down, there are hundreds, if not thousands, of news accounts of saltwater attacks detailing people killed, injured and mutilated, from Queensland to India.

Trying to get a handle on the numbers of people attacked by predators is, as we have seen, a challenge, but it is a challenge made easier for crocodiles by the existence of a database called CrocBITE.[26] This online resource is an ongoing attempt to compile all reported attacks, and is searchable by species, location, outcome (fatal, injured, unknown) and date. Doing so for 'species = CPOR' (*Crocodylus porosus* – the saltwater crocodile) and 'outcome = fatal' for a date range from 2000 to the present day brings up 892 cases. Scrolling through the first 20 reports reveals fatal attacks in 2000 and 2001 in Malaysia, Myanmar, India, Indonesia, Papua New Guinea and Sri Lanka. Where victim details are known (by no means always the case), it is the children that draw your eye. An eight-year-old girl in Indonesia, a boy of the same age in Malaysia, two 10-year-olds, two 14-year-olds. Australia may be the place with which we most associate saltwater crocodiles, and where the more attention-grabbing headlines originate, but you have to scroll down to attack number 28, a 24-year-old woman who was killed by a 4.5m (14ft 9in) crocodile in October 2002, before Australia appears on the database. On the other hand, a simple Google search for 'Australia crocodile attack' yields just under 6,000 news articles. The same simple search for Indonesia brings up just

1,200, many of which are unrelated or refer mainly to other countries or species. Australia may get the film rights and the publicity, but it is people in developing countries to the north and west that bear the brunt of saltwater crocodile predation. If we just consider fatalities during the period between the start of 2018 and the end of 2020, we get 181 reported cases of which 106 occurred in Indonesia. Just one was from Australia.

The headlines around Australian attacks are especially revealing, but you need to be thinking in the right way to spot the pattern. See if you can understand what I mean by looking at these recent headlines: 'Crocodile grabs Australian woman during night swim'; 'A swimmer escaped death after he pried his head free from the jaws of a 6-foot crocodile'; 'Swimmer bitten on head by saltwater crocodile at Lizard Island on the Great Barrier Reef'; 'Teen jumps into crocodile-infested river on dare, lives to regret it'.[27] Avoidable recreational activities in areas where crocodiles live seem to me to be a very bad idea. Compare these with headlines of attacks in Indonesia: 'Crocodile attacks and kills 55-year-old fisherman in Indonesia'; 'SWALLOWED ALIVE: Moment missing fisherman's head and limbs cut from the belly of a massive crocodile after it snatched him from riverbank'.[28] Spot the difference? Research on the victims of attacks in Indonesia found that 75 per cent of attacks occurred while they were fishing, and not participating in a recreational activity.[29] Subsistence fishing is also the highest risk in Timor-Leste, while bathing, washing and fishing are high risk in Sri Lanka.[30] In India, a similar pattern emerges, with 28 per cent of victims in a study of the Bhitarkanika National Park (just west of the Sundarbans region where tigers hold sway) fishing, and nearly a third bathing (washing) or cleaning kitchen utensils.[31] They certainly weren't night swimming or jumping into rivers for a dare.

Despite a relatively low level of human–crocodile conflict compared with other regions, we will return to Australia because a number of approaches have been developed to encourage coexistence that are potentially applicable to other species. Right now, though, we need to head back in time to the start of 1945, and to a mangrove swamp on the eastern side of the island of Ramree on the coast of Myanmar.

Mangrove terror – or wartime myth?

At the end of the Second World War, a combined British and Indian force was pushing towards what it would have called Rangoon and what is now known as Yangon, the capital of Myanmar, to liberate it from the occupying Japanese forces. Ramree Island, off the western coast, was an ideal staging post for this push, and an Allied amphibious assault force landed on the northern tip of the island on 21 January 1945. The Japanese garrison defending the island numbered around 1,000 men and, coming under sustained attack from invading forces over the following days and weeks, it eventually retreated into the mangrove swamps on the eastern side of the island. I've been in a few mangrove swamps and I have to say that if 'retreat into mangrove swamps' is your plan, then you really are in a tight spot. As a biologist, mangrove swamps are endlessly fascinating ecological and geographical wonders. Teeming with life, mangrove swamps cling on in an environment ruthlessly marked out in time by the rising and falling tide, but only vaguely defined in space as complex, slimy trees clamber over each other in a ragged boundary with the ocean. That's the scientist in me speaking; the 'human animal' in me has an entirely different opinion of the fly-laden, stinking, tangled, dangerous, mud-filled hell-hole of a mangrove swamp. I am pretty sure the Japanese garrison of Ramree didn't have much time to appreciate the ecological marvels of mangroves on the night of 19 February 1945; its plan was to get the hell out as quickly as possible and complete its retreat to the mainland. Not a bad plan, but for the flotilla of Royal Navy ships that stood in the way of its escape route – and the saltwater crocodiles.[32]

Writing in 1962, Bruce Wright,[33] a Canadian biologist serving with the British forces, vividly described the night in a book as follows:

> That night was the most horrible that any member of the M.L. [Marine Launch] crews ever experienced. The scattered rifle shots in the pitch black swamp punctured by the screams of wounded men crushed in the jaws of huge reptiles, and the blurred worrying sound of spinning crocodiles made a cacophony of hell that has rarely been duplicated on earth. At dawn the vultures arrived to clean up what the crocodiles had left. Of about one thousand Japanese soldiers that entered the swamps of Ramree, only about twenty were found alive.

Wright's account was published in a book entitled, rather sweetly, *Wildlife Sketches: Near and Far*, and Ramree rapidly gained recognition and distinction as the 'biggest man-eating orgy any crocodilians have ever been offered'. It is a story often repeated, and I remember reading about it avidly as a child obsessed with wildlife and adventure. Although it is impossible to determine who drowned and who died of gunfire, the usual story is '1,000 people eaten by crocodiles' and the 'worst-ever crocodile-predation incident', but these attention-grabbing headlines don't stand up to scrutiny. A number of people have studied the incident in depth and concluded that less dramatic factors were mostly to blame for the deaths. Foremost among them was the fact the Japanese soldiers were already in a bad state by the time they reached the mangroves, had most likely exhausted their rations and had no fresh water. Throw in dysentery and other diseases that were running rampant, and you have plenty of lethal problems without invoking a crocodilian feeding-frenzy. Local people who were present at the time have discounted the story as fantasy, while other accounts of the events of Ramree Island do not include crocodiles as anything more than a passing mention. It is quite possible that some soldiers may have fallen to crocodiles, but the story that 1,000 people were killed in the swamps of Ramree by them should be treated for what it is: a wartime myth.[34]

The fact that the Ramree 'crocodile massacre' can be accepted as fact for so long is revealing of an attitude towards predators – and an appetite for their exploits – that can be traced through real-life accounts as well as works of fiction. Even after the Ramree story has been thoroughly debunked it is still being retold in ever more lurid detail, reaching its nadir or zenith depending on your viewpoint with the release in 2021 of the film *Saltwater: The Battle for Ramree Island*, 'inspired by an astonishing true story'. However horrific, we are endlessly and viscerally fascinated by predation events.

Return to Oz

Although not the hotbed of crocodile attacks that news and films might have led us to believe, Australia does nonetheless warrant some attention when it comes to human predation risk. A study of crocodile attacks occurring in northern Australia between 1855 and June 2013 found that in the 90 years between 1855 and 1945 there were 132 fatal attacks reported (one every eight months), dropping to just six fatal attacks in the 25 years between 1946 and 1970 (one every 50 months). However, this ramped up in the 40 or so years from 1970 to the end of the study in 2013, during which there were 29 fatal attacks or one every 17.5 months.[35] These might be low numbers compared with other regions, but the northern end of Australia is relatively sparsely populated and Australia is a developed world nation. The pattern that has already become so well established, of attacks largely being focused on poorer rural communities in the developing world, is to some extent broken for saltwater crocodiles, because – as we have already seen – attack victims include recreational swimmers. However, the headlines, as usual, do not tell the full story. The data from 1971 onwards actually reveal a disproportionate number of attacks on indigenous people. In the Northern Territory, 43 per cent of all attacks and 50 per cent of all fatal attacks involved people of Aboriginal descent, against an overall proportion in the population of less than 30 per cent. What explains this bias in attacks is that a disproportionate number of indigenous people live in remote traditional homelands that coincide with saltwater crocodile habitat (60 per cent of which is on indigenous lands). Also, and now we are back with the familiar pattern, attacks on indigenous people in the Northern Territory are not related to headline-grabbing night swimming, but to activities such as fishing, hunting, gathering and generally making everyday use of natural water sources; in a very real sense this is a developing-world lifestyle within a developed world nation.

It is the rural poor rather than the metropolitan classes of Australia's cities that are most at risk from crocodile attack, but even allowing for that the number of attacks is, given the hype, surprisingly low. Another study examined attacks between 1971 and 2004 (an earlier end-date than the study mentioned above) and concluded that there had been 62 'definite unprovoked attacks by wild saltwater crocodiles, resulting in injury or death to humans'. Seventeen of these attacks caused death and 45 resulted in non-fatal injury.[36] The other study, which extended to 2013, concluded that 29 people had been killed since 1971 and – while these were unimaginably horrible deaths both for the victims and their families – these numbers do not reflect the common notion that Australia is teeming with lethal crocs. However, attacks do seem largely to be motivated by predation (90 per cent of fatal and non-fatal attacks between 1971 and 2013) and crocodile numbers are rising, which is leading to increases in attacks. This rise in crocodile numbers, and the reasons for it, is a conservation good news story – and it is a story that tells us much about the pragmatic resolution of emerging human–wildlife conflicts.

Between 1946 and the early to mid-1970s, northern Australian saltwater crocodiles were targeted by unregulated commercial hunters for their skins. The skins have smaller scales than those of other species, making their patterns look better for more delicate items, and the lack of osteoderms (small bones within the skin) make the skins easier to work. These qualities made saltwater crocodile skins valuable to the fashion industry and valuable wildlife products tend to attract heavy commercial exploitation. There was money to be made and the first 10 years or so after 1946 saw the resulting hunting frenzy peak. By the time the 1950s and early 1960s came around there was an inevitable decline of saltwater crocodiles (and in their attacks, see previously), which caused a shift in hunting focus to freshwater crocodiles (sometimes known as 'freshies') with inferior skins. Individual states acted to protect the reptiles, with formal legal protection in Western Australia coming in 1969, the Northern Territory in 1971 and Queensland in 1974. By that point it has been estimated that saltwater crocodiles had been reduced to less than 5 per cent of their historical population and less than 1 per cent of their biomass, the largest animals having been selectively hunted for their larger, more valuable skins. Once afforded legal protection, saltwater crocodiles were able to recover. Such is the success of the programme that it is probable crocodiles have reached close to their pre-hunting levels in abundance and biomass, and are nearing their carrying capacity (the maximum number that the environment can support).[37]

Crocodiles rebounded relatively quickly following legal protection and even by the 1980s the population had risen from 5,000 to around 30,000. That sixfold increase was more than enough to put crocodiles into direct conflict with humans. A series of fatal and non-fatal attacks around this time led to calls for the recovery programme to be ceased and widespread culling of the animals was actively promoted. In other words, within a decade the predators had rebounded to such an extent that people now wanted them controlled. This sets the stage for a classic human–predator balancing problem. On the one hand, driven by conservation and biodiversity goals, we want to build up populations. On the other hand, we want to be safe from predators. The progression from over-exploitation to near-catastrophic decline, to conservation intervention to conservation success to increased conflict with humans and then calls for culling needs to be broken early, before increasing numbers become a problem. One way to enhance coexistence is to find a means to pivot the predator from problem to opportunity. This is exactly what happened in the Northern Territory.

If there is a dangerous animal roaming around then it is perfectly natural to want to remove it. You may not wish the animal dead, you might desire another solution, but overall it is normal to want to live in a safe environment. In general, even with good intentions, the human–predator balance tends to topple over towards the human side because the risk of having predators around simply outweighs the benefits. There are really

only two ways to shift towards a more favourable
balance for predators. One way is to reduce the risk to
humans without exterminating predators by mitigating
against negative interactions through the sort of
interventions already detailed: predator-proof *bomas*;
improved dwelling construction; human behavioural
changes and so on. The other way, which can occur at
the same time as reducing risk, is to increase the benefit
that people gain from predators. The best way to do that
is to make predators valuable by providing people with
incentives to coexist with them. That can be very
difficult in practice because we tend to value our safety
very highly. More dangerous predators require far
greater incentives to keep them around since the risk
that must be defrayed by benefits accrued is far greater.
In the case of saltwater crocodiles, there are currently
two ways that people can gain financially from their
presence: tourism and skins.

Tourism is reasonably easy to understand from a
conservation perspective. In the ideal world, wildlife
tourism provides a range of jobs and income-generating
opportunities that rely on having wildlife present, and
thus incentivise people to maintain wildlife and the
habitat it depends on. It is not always an ideal world,
though, and tourism is sometimes linked with
corruption, exploitation and other issues, including
degradation of the habitat on which it depends and a
high carbon footprint. However, those problems
notwithstanding, you need certain things to be in place
for tourism to work. First, you need a reason for people
to visit an area; you need something that will draw them
in. Then, you need reasons for them to stay for enough

time to spend their money locally. Finally, you need the infrastructure and skills to support the industry. Saltwater crocodiles are most certainly a big draw; people want to see them and are willing to pay money to do so. The attraction of crocodiles has proved to be enough to develop the infrastructure needed to have a tourist industry. What has resulted is a wide range of crocodile-linked activities, from captive encounters to 'crocodile safaris', which offer people unique wildlife experiences and bring money into more remote areas with few other resources to trade off. The industry is worth tens of millions of dollars annually, which may not seem like much when compared with the GDP of Australia, but it is most certainly a significant income for the people involved. And, when your prosperity relies on the presence of crocodiles you are going to make sure there are plenty of crocodiles around for people to see.

Seeing live crocodiles as a touristic experience has solid economic value but, as we've seen, dead saltwater crocodile skins are highly desirable and valuable. Of course, you may not think that. You may find crocodile leather shoes, bags, wallets, belts and jackets hideous, but many people do not. It is, after all, the desirability and value of their skins that led to the unregulated and nearly catastrophic hunting of the species between 1946 and its eventual protected status in the early to mid-1970s (depending on the state). If you are the sort of person who finds wearing or carrying accessories made from animal skins sickening then prepare to be sickened, because the conservation of saltwater crocodiles in Australia and their amazing rebound depended – and still depends to a great extent – not on the joy of seeing one alive but on the commercial value of their skins once they are dead.

In the 1980s, the Northern Territory government started an incentive-driven conservation scheme that explicitly recognised – and took advantage of – the commercial value of saltwater crocodile skins. Positive incentives were created for people through the establishment of crocodile farming and ranching, driven by sustainable harvesting programmes focused on egg collecting. Rural communities harvest saltwater crocodile eggs from the lands under their control. These eggs are then incubated and reared in captivity on commercial facilities until the age of around three. By this time the crocodiles are about 1.5–1.8m (4ft 11in–5ft 11in) long and are big enough to be killed and converted into leather.[38] There is no point in dressing up the fact that the industry exists to kill crocodiles and to make leather for non-essential fashion items. It does not exist to conserve crocodiles, but neither would it exist if crocodiles were wiped out. Hermès and Louis Vuitton are the big players here, and they don't tend to make a big noise about their involvement for obvious reasons.[39] However, because of their involvement, the people collecting eggs get around US$14–28 per egg. With more than 60,000 eggs collected from Aboriginal land, that is a decent amount of money coming into communities that are otherwise far from prosperous. The incentives (cash) provided by this sustainable harvest have been key in developing public and political tolerance for saltwater crocodile reestablishment, and have led to crocodile habitat being highly valued by landowners and managers. The extensive wetland areas where crocodiles (and therefore their eggs) can be found are now protected and encouraged, with many other species benefiting. It makes sense. If you have a rat in your house, you might be tempted to set a trap out. If the rat is worth £10,000 then you might be tempted to attract more rats.

Lest we get too rosy-tinted in our view of crocodile utilisation it is worth noting that the overall value of the crocodile-skin trade is estimated to be around US$75 million per year. This is far more than people are willing to pay to see them alive, sadly, but it is also far more than finds its way into the Aboriginal communities whose lands and labours support the industry. It is also very easy to fall into a logical trap here. It would be easy to suggest that the economic value of crocodile skins is what saved the species, while forgetting that this exact same value is what threatened them in the first place. For me, the key take-home messages of the saltwater crocodile story are that the commercial utilisation of a species can provide huge conservation benefits and that financial gains can be sufficient to offset the risks posed by predators, but only when the utilisation is sustainable and when the distribution of benefits is fair.

There is a third way in which saltwater crocodiles could provide sufficient financial benefit in principle to offset the risks they pose: trophy hunting. Hunting is a recreational activity for many people around the world and for many the motives that lead them to hunt may be complex. People may want to acquire meat, they may be seeking an experience, they may want to be more involved with the process by which food finds its way on to their plate, or they may be driven by a desire to hunt a particularly large or otherwise notable individual animal. In practice these motivations, and others, are rarely mutually exclusive, but it is the final motivation, the desire to hunt a particularly notable specimen and to keep some part of that animal, which falls into the realm of trophy hunting. The 'trophy' may be the horns, the skull, the skin or some other part of the animal, which acts as a memento of the hunt. If you've ever been to a luxury hotel or a large country house then you've probably sat in a room with a stag's head on display, while a visit to pretty much any antiques shop could secure you a deer or other animal head for your own home if you wanted. Pretty much any tourist accommodation in southern Africa also comes complete with a kudu skull and horns, or a buffalo head. I guess what I am saying is that putting heads on walls isn't really unusual, but few people have the money or the desire to secure one for themselves. Those who do have the means and motive to do so can find plenty of opportunities around the world to hunt a variety of species. As is the case with fashion houses, their motivation is not conservation but, by paying to hunt, they can provide a financial incentive for landowners or communities to conserve habitat and wildlife on land that might otherwise provide more return if converted to farmland. When this system works well it can be very effective, as evidenced by the recovery of wildlife in countries like South Africa and Namibia. However, it is a complex world and a complex topic, and returns don't always make their way to the right people. Nonetheless, in principle, and very often in practice, this type of tourism (for it is largely overseas hunters that we are talking about) can provide sufficient benefit to offset the costs imposed by certain more problematic species, including predators.

If you want to hunt crocodiles, then currently the Nile crocodile is the species of choice, with hunts available (at least at the time of writing) in South Africa, Zimbabwe, Mozambique, Tanzania, Zambia, Namibia, Uganda and Ethiopia. Saltwater crocodiles are not currently a species that can be hunted, but it has been suggested more than once that they should be. In 2015, before the Cecil the lion incident, it was reported that Australia was considering allowing the hunting of salties as a way to bring income to poorer communities.[40] The indigenous affairs minister Nigel Scullion suggested that hunters would pay £15,000 to hunt larger individuals. Further calls came in 2016, this time motivated by removing individual crocodiles that posed a risk to human safety.[41] Calls to cull saltwater crocodiles to keep their numbers in check are also regularly made, but so far culling has not received political support.[42] Adam Britton, who keeps a large saltwater crocodile in his back garden, as I found out when I interviewed him in 2021 for a BBC Radio documentary on crocodiles,[43] has spoken out against a culling program, stating that 'the easiest way to keep people safe is to make sure they understand the risks' and to change their behaviour around crocodiles to reduce the risk of attack, for example by not getting into the water with them.[44] Professor Grahame Webb has also discussed the problems with culling, including deciding what level is 'safe', deciding where to cull and the potential for culling to, counterintuitively, end up with more crocodiles by removing the larger individuals that may be controlling the overall population. He concludes that 'selective culling has a role to play in the overall management of crocodiles, but is not the public safety panacea that it may superficially appear to be.'[45]

At the moment, crocodiles are safe from hunters, but if culling were approved then to me it makes little sense not to sell at least some of the cull quota to hunters willing to pay. That way, culling is turned from a cost into a potentially large benefit – and it makes no difference to the crocodile who has pulled the trigger or what their motivations are. As long as the process is well-regulated and humane, I believe it is up to Australia – and particularly those communities that live with saltwater crocodiles and have done such a good job in conserving them – to decide. However, I suspect that pressure from largely urban-based animal-rights groups – and groups that live on the other side of the world – may well prevent this form of benefit from being realised even if it were decided that culling were an appropriate course of action.

Mugged

The saltwater crocodile, though present in India, is limited to coastal areas. Away from the coasts, in inland waterways, there is another species that also has a fearsome reputation when it comes to eating people: the mugger crocodile. Found in southern Iran, Pakistan, Nepal, India and Sri Lanka, and possibly in Bangladesh, Bhutan and Myanmar, the mugger is a medium-sized crocodile, with big males typically reaching around 3–3.5m (9ft 10in–11ft 6in) and only rarely exceeding 5m (16ft 5in). The species has adapted well to more human-dominated landscapes and is frequently found in reservoirs, irrigation canals and artificial water bodies. It may also have some overlaps with saltwater crocodiles, being found at times in coastal lagoons and estuaries. From a conservation perspective mugger crocodiles aren't doing so well, with habitat loss and fragmentation being a big factor in likely declines that have led to some estimates of their numbers in India being put at just a little over 4,000.[46] They were hunted illegally for the skin trade that affected the saltwater crocodile (*i.e.* in the 1950s and 1960s), but changes to habitat and getting trapped in fishing nets are the major threats to the species currently. The species is legally protected through much of its range, and habitat protection and restoration is clearly a priority. Some restocking programmes using captive-bred individuals are under way and, with an eye to the successes in Australia (and similar successful programmes in parts of Africa), sustainable utilisation is being mooted.[47] However, sustainable utilisation of mugger crocodiles, including for tourism, can only work if there is a market for them, either as a touristic experience or for skins and meat. At the moment, that is not really the case and crocodile farms set up for visitors are struggling.[48]

Given that mugger crocodiles are far from abundant, particularly in comparison with the Nile crocodile through much of its range, it is notable that it accounts for a fair number of human victims. Between 2010 and 2020, 174 people are listed on CrocBITE as having been killed by them, and extending the search to include non-fatal encounters brings the number up to 417. In fact, muggers are the third most dangerous species to humans and most attacks occur in India, where the majority of muggers are to be found. Where muggers coexist with saltwater crocodiles in Sri Lanka the data show that muggers are far from the least aggressive species.[49] An analysis of attacks between 2010 and 2015 found that salties killed 27 people, but muggers managed nearly double this, killing 49 people in the same period. More than three-quarters of victims were male (77 per cent) and most were killed, yup you guessed it, bathing, washing clothes and collecting resources near the water's edge. Whether it's Nile crocs in Africa, muggers in India or salties in Indonesia, the picture is much the same.

Crocodilians are remarkable predators. Evolution has shaped their form and general ecology into a group of species that do very well in a range of aquatic environments. As a consequence, even with human exploitation and widespread habitat destruction, they remain, as a group, relatively common throughout much of the tropics. That these areas frequently coincide with places where humans live, and that both humans and crocodiles have a focus on water, results in encounters being inevitable. However, it is not inevitable that such encounters have to end poorly for either us or them. Unlike attacks by other predators, crocodile attacks occur at reasonably predictable general locations, principally in or very near to water that contains them. Mitigating against the risk of attack is not notionally that difficult but, as we have seen, it requires a level of development, prosperity and infrastructure organisation that may still be beyond the reach of many people who are most at risk. Relying on water bodies for everyday needs, from washing through to food and building materials (such as reeds) exposes people to predation risks that, in some places, may be quite severe. It is worth remembering that donating money to build better wells or fund solar-powered pumps doesn't just provide people with fresh, clean water. By removing from their lives a reliance on visiting local rivers or water bodies you may well be saving families from the horror of a crocodile attack. Development isn't a dirty word, and foreign aid isn't just about sacks of rice.

CHAPTER FIVE

Forest Legends

So far, I've focused on species that undeniably regard us as prey and whose predatory activities can cause considerable issues for communities that live alongside them. I'll return to that theme in further chapters where I'll evaluate the dangers of other large predators, including leopards, bears and wolves. But, for now, I'd like to take a diversion and consider a trio of rather less obvious potential human-eaters, all of which have more than a whiff of myth and legend about them.

A six-legged army

Most people have watched, fascinated, as ants follow invisible trails up a wall, over a floor or across a kitchen worktop. Ants have a remarkable system of foraging based around chemicals collectively known as trail pheromones. Foragers leaving the nest dribble a line of trail pheromone behind them that other ants are able to follow. If these foragers find food then they can follow their own trail back to the nest, while also reinforcing it with more pheromone. The more pheromone is laid down on a trail the more likely other ants are to follow that trail. When combined with diffusion and evaporation of the trail pheromone – and other pheromones that further tweak the ants' behaviour – ants are able to use trail pheromones to collect food across a complex landscape without the need for any central control. Ants don't have bosses telling them what to do, yet nevertheless their activities appear to be the product of some centralised intelligence. Highly efficient and effective, ant collective behaviour is the inevitable product of simple rule following.

My early scientific career focused on ants, especially the behaviour of a group of species called the leafcutter ants. These are the ants you see in documentaries carrying fragments of leaves along the forest floor in fluttering green 'conveyor belts'. These forager ants are heading back to their nest where other ants will break the leaves down and use them to nurture a fungus that is only found in the nests of leafcutter ants. It is this 'farmed' fungus that the ants use to feed their larvae. Leafcutter ants are endlessly fascinating and I've spent close to 25 years studying them one way or another. That said, the first time I saw these ants in the wild – cutting leaves in a tropical forest in southern Mexico – is a strong memory for me because a far more spectacular piece of ant biology was going on nearby. Across the forest floor, a short distance from the green highway of ants that had first attracted my attention, thousands of army ants were on the move.

I'd grown up reading fantastical tales of ferocious army ants and seeing a large colony of them on the move through the forest for the first time was a thrill. Tens of thousands of pale-headed ants, some bearing spectacular sabre-like jaws, were carving up anything that was unlucky enough to be in their way. Sometimes whole spiders and insects were being ferried back, but at other times the ants had stumbled across something too big to carry without some forest butchery – a still-twitching spider's leg particularly sticks in my mind. I've seen army ants in Central and South America many times since, and they are enthralling to watch, but they also have quite a reputation. A good demonstration of this is the fact that typing 'Can army ants' into Google returns the suggestions 'kill a human', 'kill an elephant' and 'kill you'. To find out whether their fearsome reputation and potential man-eating status is justified first requires us to explore what 'army ants' are, what they are not and how they operate.

Foraging-trail pheromones and self-organised behaviour mean that many ant species are able to gather food collectively. However, in reality that collective action is only apparent when we 'zoom back' and consider ants at the level of the ant nest or the colony. Several hundred ants trooping across the ground to forage on a dead mouse (a scenario that is playing out in my back garden as I type this) is a collective action in terms of getting food back to the colony, but individually each ant is very much fending for itself. Each individual is 'doing its bit' for the greater good, but there isn't much cooperation going on at the dead mouse. In fact, if you watch the ants carefully, you can often see them getting in each other's way or fighting over individual morsels of meat. I see the same thing in my laboratory colonies of leafcutter ants, where individual ants will sometimes battle over a piece of leaf, tussling for several minutes before one, usually the largest, wins.

The 'army ant' species have evolved a form of collective behaviour that is far more coordinated. Known properly as 'legionary ants', ants of these 200 or so species distributed around the world collectively embark on huge, highly coordinated predatory 'raids'. The entire colony leaves the nest, sweeping through the environment, grabbing anything that is unable to get out of its way. Working together, the ants use formidable jaws to slice up prey and ferry it back to the hungry larvae. In some species, individuals work together to carry tricky prey items, with a larger ant acting as the main traction 'engine' while a smaller ant helps to steer and direct the prey in the right direction. Other studies have shown teams developing, with specific ants electing to work with each other repeatedly, with associated efficiency gains; team work makes the 'carve up everything in your path' dream work. The nests of these ants are usually very different from the sort of nest you might find under a rock in the garden. Legionary ants are able to forage so effectively that they rapidly deplete their surroundings of prey. This requires them to periodically shift to new hunting grounds. Consequently, their nests are temporary arrangements, often termed bivouacs, with ants forming the structure from their own bodies and keeping the queen and brood (eggs, larvae and pupae) safe in the middle. They are tiny but remarkable predators.

The legionary ants of Central and South America, especially those in the genus *Eciton* (though there are a few others) are the species we call army ants, while the 70 or so species in the genus *Dorylus*, found in Africa, are often called driver or safari ants. With a similar overall lifestyle to the army ants, the driver ants of Africa have ramped everything up a notch and are unarguably more spectacular. They can have enormous colonies, potentially reaching 20 million or more individuals, and their raiding columns are flanked by huge soldier ants sporting immensely powerful jaws. Driver ants are highly organised, with a fearsome reputation, massive strength of numbers and powerful jaws. This all leads to an inevitable question: can we add some ant species to the growing list of animals that kill and eat us?

Terror of the forests?
The first thing to say is that if you are mobile then you have nothing to fear from these ants. If you are on their trail then they would swarm all over you if you gave them the chance, but it is perfectly possible to simply take a step to one side and avoid them completely. Some army ants have a broad trail 'front' that sweeps out from the main trail like a wide river delta; these more diffuse trail systems are harder to avoid, but you can still get out of their way very easily. You may have seen breathless tales of how fast ants move, but these incredible figures are based on 'scaling up an ant to the size of a human'. In real life ants are the size of ants and even the relatively swift army ants can only move at perhaps 5cm (2in) a second or so. Given that army ants announce their presence on your body by stinging and biting, it is safe to say that you'll be keen to move out of harm's way very quickly and certainly before you are overwhelmed.

If you couldn't get out of the way of these ants, though, things could turn nasty. I was chatting to a smallholder in Mexico once who was tending his beehives. Each hive was arranged on homemade concrete troughs that had been filled with old engine oil. When I asked him why, the answer was '*hormigas armadas*' – army ants.

He then nodded towards an abandoned chicken run. Army ants regularly swept through the area and, like driver ants in Africa, can provide a very useful pest-removal service in houses, outbuildings and crops. However, the defences on the chicken run had not held. The ants had managed to get in and the chickens had not managed to get out, with predictably grisly results. I have heard similar stories of driver ants in Africa overwhelming tethered goats and even chained crocodiles.

Marauding ants have long been a staple for writers and filmmakers. There is something undeniably horrible about the prospect of being overrun and consumed by insects, and when you throw in the apparent intelligence and relentlessness of ants then you have perfect 'villains'. There is even an episode of American action-adventure series *MacGyver* that has army ants as the protagonists, the ever-improvising hero escaping with the usual suspension of the rules of physics and chemistry. That episode, 'Trumbo's World', has a fascinating ancestry that reaches back to some of the earliest fictional writing on army ants. 'Trumbo's World' is heavily influenced by the Charlton Heston film *The Naked Jungle* released in 1954, in which Heston plays cocoa-plantation owner Christopher Leiningen. The film is largely based around the romantic tension between Leiningen and Joanna (played by Eleanor Parker), but it ends in dramatic fashion when Leiningen has to blow up a dam to thwart an advancing raid of army ants. *The Naked Jungle* was itself based on a short story called *Leiningen Versus the Ants* by Carl Stephenson, an Austrian-born German writer. The story gained some notoriety after appearing in *Esquire* magazine in 1938. Radio versions were made in 1948, and again in the late 1950s, after the release of *The Naked Jungle*. It was most recently revived as a *15-minute Drama* on BBC Radio 4 in 2018, some 80 years after the original story first appeared. Villainous ants clearly have long-lasting appeal.

Leiningen Versus the Ants may have proved fertile ground for dramatic adaptations, but as a biological work it is a little overblown.[1] Telling Leiningen of the impending 'invasion', a Brazilian official, throwing up his 'lean and lanky arms with wildly distended fingers', pulls no punches when he tells the plantation owner that he's 'insane! They're not creatures you can fight – they're an elemental – an "act of God!"'. Describing a column of ants 10 miles (16km) long and 2 miles (3.2km) wide, the official goes on to explain how they can 'eat a buffalo to its bones in seconds'. As vivid a description as that is, a column of ants 10 miles (16km) long and 2 miles (3.2km) wide is as fantastical as the notion that ants can strip a buffalo to a skeleton in seconds. Some later descriptions of the ants are a little more realistic, including their ability to build bridges over obstacles with their own bodies, but overall the fictional horror far exceeds the real-world threat.

Fictional flourishes and obvious plot holes in dramatic depictions certainly do not remove the possibility that army and driver ants *could* kill and eat a person. In support of this possibility is the fact that there are credible reports of people being executed by ants. The French-American traveller, zoologist and anthropologist Paul Du Chaillu travelled extensively through West

Africa in the 1850s, famously becoming the first modern European to see gorillas. Du Chaillu mentions driver ants, referring to them as 'the bashikouay'. Among some solid descriptions of their general biology, interspersed with the obligatory 'skeletonising' of a large mammal or two, he mentions that 'criminals were in former times exposed in the path of the bashikouay ants, as the most cruel manner of putting to death'.[2] Cruel indeed, but also unusual.

When it comes to less deliberate deaths, I have long heard rumours in Africa that driver ants have killed people. Typically, the stories involve infants, elderly or inebriated people being overwhelmed and suffocating to death as ants invade the nose and mouth and consume whatever flesh they can get hold of. William Gotwald, in a 1980s article evocatively entitled 'Death on the march', describes hearing stories of driver ants killing unattended babies from a village chief. Driver ants were also apparently implicated in the death of a missing tourist in Tanzania in the early 2000s,[3] but I've been unable to find any solid confirmation of this. I've asked a great many people with considerable combined experience of living and working in the African bush about driver ants killing and eating people, and it is always the same story: rumours of deaths, especially of infants or drunk people, but never anything concrete. However, there is a medical report of a man who was extensively bitten by driver ants that certainly suggests that the rumours can be, to some extent, believed.

Published in 2010 in the *American Journal of Tropical Medicine and Hygiene*, the report tells the story of a 40-year-old man admitted to an emergency department of a rural hospital in Uganda.[4] The man, with a history of alcohol abuse, 'had spent the preceding night sleeping in the bush'. This is as a good a euphemism as any for 'had collapsed blind drunk', which is certainly the implication of the article. The man had been attacked by driver ants (called safari ants by the authors) and was covered with 'multiple erosions' all over his body, caused by ant bites. The image that accompanies the paper is fairly dramatic and it is clearly possible to see that the multiple bites had resulted in the removal of flesh. The man was discharged after five days of care, but his story shows us that if someone is sufficiently immobilised, for whatever reason, it is clearly possible for driver ants to consume at least part of them. Were there enough ants to invade the nose and mouth then suffocation could result. Though uncommon, and poorly documented, I am confident that we can add driver ants to the roster of animals that could kill and eat us as an act of predation. However, looking at the photograph of the man in Uganda, and the wounds he suffered, it seems more accurate to think of ant predation on immobile and likely unconscious people as a form of scavenging. Either way, it is a fate best avoided. There is a phrase I first heard in Malawi, but is doubtlessly used all over the world: 'one for the ditch'. The ditch drink is the final 'final' drink, after you've already had 'one for the road'. My advice is to pass on the ditch drink and avoid bush naps.

Attack from above

Driver ants might well nibble you if you're immobile on the ground, but they don't really fit the profile of human-eating predators I established back in Chapter

One. They manage to bypass the requirement to be relatively large by being incredibly numerous and well organised, but most people can easily get out of their way. It is much harder to get out of the way of a predator that flies, but you might find it hard to believe that any bird would consider a person as prey. After all, even birds of prey are relatively small and not especially robust when compared with many other predators. I certainly thought that, before seeing a harpy eagle in the flesh.

Harpy eagles are huge. Females are larger than males, which is a common pattern in birds and other egg-laying animals including reptiles, fish and insects. An average-sized female might weigh in at around 9kg (20lb), although a captive female named Jezebel was over 12kg (26lb). They can stand more than 1m (3ft 3in) tall and have a wingspan in excess of 2m (6ft 7in). If, like me, you find it hard to picture a bird from a series of dimensions then let me tell you about the first – and indeed only – time I saw a living harpy eagle. I was a Fellow for a few months at the Smithsonian Tropical Research Institute in Panama, based in the tropical forests around the Panama Canal. Keen to see a harpy eagle, the national bird of Panama, I went to an aviary where they were trying to breed the birds for a release programme. I can remember quite clearly seeing a large female eagle. I was struck by her huge legs and thick, long talons. I also recall the person I was with saying that it was 'a scary, no-nonsense bird', but sitting there on a perch barely moving and seemingly oblivious of the world around her she didn't seem too threatening. Then a toddler came around the corner, a lively child somewhere between two and three. The eagle visibly stiffened. Its entire demeanour changed in a heartbeat. It swivelled its head towards the child, leaned forward and, like a switch had been thrown, turned from almost asleep to hyperalert. I was with two other people and we all had the same goosebump feeling. The eagle had seen prey.

That harpy eagle may have perked up at the sight of a small child, but grown adults had no effect on it whatsoever. This is hardly surprising, because adults are well outside the realistic size range for prey of even the very largest eagle species. There is only a handful of truly large eagles in the world: the harpy, the Philippines eagle, the African martial eagle, the African crowned eagle, Steller's sea eagle and the white-tailed eagle. Each of these massive raptors is certainly impressive and it is clearly possible for a large eagle to swoop down and embed its massive talons in your skull. This would very probably be a fatal wound. With its beak and talons an eagle could then eat portions of you while you were dead on the ground. From that point, it would be a trivial exercise for the eagle to take some morsels back to its chicks. In principle all this is possible, but there are just no credible reports of this happening to adults. Children are a different story.

Harpy eagles can pluck black howler monkeys directly from the tree canopy. Howlers are the largest monkeys in Latin America, with males reaching 60–65cm (2ft–2ft 2in) in length excluding a long tail and 12–14kg (26–31lb) in weight. That is in the same ballpark as an 18-month-old or small two-year-old. A human toddler is certainly a possible prey item, but despite this there seems to be no credible report of a harpy eagle taking a child. A factor in this might be the way harpy eagles hunt. A massive eagle adapted to dense tropical forest canopy hunting might find it difficult to take a child at ground level. On the other hand, eagles adapted for hunting ground-based prey in more open country, or in woodlands, might find a child easier to take. It is to these species we must look if we want to find out whether eagles predate children.

The crowned and martial eagles of Africa are good candidates. Both species are large, with the martial eagle being the larger. I've seen both species and, in the flesh, martial eagles are comfortably the more impressive of the two. Height- and weight-wise, though, neither African species is as large as the harpy eagle, although the martial eagle has a larger wingspan than its Latin American cousin. Despite being smaller and less imposing than the martial eagle, it is the crowned eagle that is considered the more powerful in terms of the prey it is able to tackle. Crowned eagles have been known to hunt bushbuck (a small species of antelope) and in some parts of their range they frequently take large monkeys of a not dissimilar size to a toddler.[5] The big difference between these eagle species and the harpy eagle is that the African species tend to live in far less densely forested areas, and they will take prey at ground level far more often. Harpy eagles can and will take prey on the ground, but they tend not to. One study found that 79 per cent of all prey was sloths taken from trees, and tree-living monkeys make up most of the rest. The majority of harpy eagle prey thus weighs 2–3kg (4–7lb) and lives in the top branches of dense tropical forests. On the other hand, martial eagles and crowned eagles have diets that include ground-living mammals larger than a toddler including duikers (a 15kg/33lb antelope), Thomson's gazelle and impala.[6]

It is the more powerful crowned eagle, with its dietary penchant for primates, which seems the most likely species to target humans. There is some evidence to support the idea that this happens, albeit very rarely. The most definitive reported case is described in Peter Steyn's *Birds of Prey of Southern Africa: their identification and life histories*. According to the author, a seven-year-old boy was attacked and 'savagely clawed' on the head, arms and chest by an immature crowned eagle, and was only saved by the quick actions of a woman brandishing a hoe.[7] The eagle did not survive, but the boy, Damas Kambole, was successfully treated at a nearby mission hospital.[8] The incident occurred far from any known nest and the bird was immature, which are important facts for interpreting what occurred. Eagles, like many birds, can be highly defensive of their nests, especially if chicks are present, but for an eagle to attack a 20kg (44lb) boy far from a nest strongly suggests predatory intent. There is other evidence of crowned eagles taking children. Raptor conservationist Simon Thomsett describes a 'macabre anecdote' while 'investigating an alleged kill of a human infant (four-year-old girl)'. He was 'brought to the tree where her severed limb was found. The circumstances led to no doubt that the accusation [that the child was taken by an eagle] was true, for no leopard could have climbed that tree.'[9] A portion of a child's skull has also been found in the nest of a pair of crowned eagles, although in that case scavenging could not be ruled out.[10]

The larger but less powerful martial eagle has also proved capable of attacking and killing human children. In 2019 in the district of Gaashaamo in the Somali region of Ethiopia, a child was killed by a martial eagle that also wounded at least two more children.[11] The eagle was subsequently killed, with some reports suggesting it was shot, but others reporting that a teenager killed it with a cane after the eagle had attacked him. Whether these attacks were defensive or predatory in intent is impossible to determine, but they nonetheless clearly demonstrate two facts: eagles can and sometimes do attack people, and they can kill us.

There is some tantalising evidence that eagle predation may have been a more important factor in our evolutionary past, when our ancestors were smaller. Also, as we'll see shortly, some extinct species of eagles were considerably larger than modern-day species. The Taung Child is a fossilised skull of an *Australopithecus africanus*, a species of hominin that lived in South Africa between 2 and 3.7 million years ago. The Taung Child, was found in 1924 by quarrymen working in a limestone quarry near the small town of Taung in North West province, South Africa. The town's name means 'place of the lion', the Tswana word for lion being *tau*, but it is a different predator that has become inextricably linked with the Taung Child. The history of the Taung Child fossil is a fascinating story in itself, with the Piltdown Man hoax casting a shadow over initial claims that the skull represented a human ancestor. Eventually, over the course of several decades, the skull was accepted as being that of *Australopithecus africanus* and is now considered one of the most important hominin fossils. Indeed, the Taung

Child is still intensely studied close to a century after its discovery. One such study, published in 2006, concentrated on the possible cause of death. This is where an eagle enters the story.

Lee Berger is an American-born paleoanthropologist in South Africa, as well-known for his discoveries as for his high public profile. He is perhaps best known for leading the expedition that found another hominin, *Homo naledi*, in the Rising Star Cave system in South Africa, a discovery that was published in 2015. Before that discovery, he become well known for advancing the Taung bird-of-prey hypothesis. In 2006, he published a short paper that compared the skull damage of monkeys that had been killed by crowned eagles with some distinctive damage discernible in the Taung Child's skull.[12] Berger found that, 'Re-examination of the Taung juvenile hominin specimen ... reveals previously undescribed damage to the orbital floors [the base of the eye socket] that is nearly identical to that seen in the crania [skulls] of monkeys preyed upon by crowned hawk eagles [an alternative common name for the crowned eagle].' Other non-hominin bones found with the Taung Child lend weight to the idea.[13] The assemblage of bones is typical of the accumulation of prey species' remains associated with birds of prey. Taken together, the evidence strongly points to the Taung Child having been taken by an eagle.

The Taung Child was small. Estimated at just over 1m (3ft 3in) tall and weighing around 10kg (22lb), they would have been the size of a small two-year-old modern human, or more realistically, perhaps a large 18-month-old toddler. Taller, but lighter, than a howler monkey from South America, and a little heavier than a typical vervet monkey from Africa, the Taung Child is certainly in the prey size spectrum of modern eagles, but arguably only just. Human size is of course only one side of the eagle–human equation. The other way to push humans into the prey spectrum of eagles is to make the birds larger.

The biggest eagle of all

No account of eagles hunting humans would be complete without mention of the largest eagle known to have existed, the mighty Haast's eagle. Estimated to have weighed 15–18kg (33–40lb) – comfortably 50 per cent heavier than the harpy eagle, and perhaps as much as 100 per cent heavier than a typical individual – Haast's eagle was native to the South Island of New Zealand and went extinct around 1400. Its large size has been explained as an evolutionary adaption to hunting moa, the huge flightless birds that used to roam New Zealand, now also sadly extinct. Looking like an oversized emu, the largest species of moa were more than 3m (9ft 11in) tall and weighed in excess of 200kg (441lb). To explain Haast's eagle's large size we need to explain the moa's large size – and to do that requires a small detour into some evolutionary theory.

The huge size of New Zealand moas is attributed to an evolutionary process termed 'island gigantism', whereby small species that end up on islands tend to increase in body size. One reason for this evolutionary increase in size is that large mammalian predators are often absent from islands. It is much harder for a big predator to make it to an island. Being small, and able to escape and hide, is a good adaptation to counter predators, but if predators are absent then prey is released from this constraint. Colonising islands is not easy, so those that manage it may not have much competition from other species. This lack of competition and absence of predation can favour larger, slower-growing individuals, a pressure that over time can lead to exaggerated size and, in birds, a loss of flight. The dodo, essentially a giant pigeon, is a good example, as are the elephant birds of Madagascar. Similar to the moas, elephant birds were generally more heavy set; the largest species, *Vorombe titan*, was slightly shorter than the largest moa but close to three times its weight. The evolutionary flipside to island gigantism is 'island dwarfism'. In this case, evolutionary pressures push animals in the opposite direction and favour small size. One explanation is that if food resources are limited then conditions favour animals that mature earlier at a smaller size. Individuals that are smaller and breed sooner produce more offspring over time than larger, later-breeding individuals and so smaller forms come to dominate. One well-known extinct example of island dwarfism is an elephant that lived on Malta and Sicily. This species stood less than 1m (3ft 3in) high, and weighed less than 300kg (661lb). It is thought that the skulls of these tiny elephants may have led to the legends of Cyclops in Greek and Roman mythology. Elephant skulls are distinct in having a large central nasal cavity accommodating the trunk, which appears as a big hole in the middle of the front of the skull and could be mistaken for a single large eye socket.

But back to pre-1400 New Zealand and Haast's eagle. CAT scans of what remains we have of these birds were used to study their evolution and to infer features of their lifestyle and ecology.[14] The only extant species of birds that are meaningful comparators to the Haast's eagle are vultures and condors. Both of the largest species, the black vulture (around 14kg/31lb) and the Andean condor (around 15kg/33lb), are a little smaller than the largest estimates we have for Haast's eagle, but they are in the right ballpark. Both of these birds are predominately scavengers and, because its beak was similar to the beaks of vultures, it has been suggested that Haast's eagle had a similar scavenging lifestyle. However, a detailed study of bones, especially the pelvis, suggest that the bird was more than capable of delivering a fatal blow when swooping on prey. The study provided what palaeontologist Trevor Worthy described in an interview on ABC News as, 'convincing data [showing] beyond doubt that this bird was an active predator, no mere scavenger.'[15]

We already know that extant eagles can potentially kill small humans and, having established the credibility of Haast's eagle as a predator, the size of the bird clearly represented a threat to humans. The first Māori arrived in New Zealand South Island around 1300 and it is their hunting of moa that likely precipitated the relatively rapid extinction of Haast's eagle. However, between early colonisation of potential prey and the eventual extinction of predator there was plenty of time for interactions, some of which may have made their way into mythology. Māori stories include the *pouakai* (or *poukai*), a massive bird that killed and ate people, the origin of which may have been Haast's eagle. As one of the authors of the bone study, Paul Scofield, puts it, 'The science supports Māori mythology of the legendary *pouakai* or *hokioi*, a huge bird that could swoop down on people in the mountains and was capable of killing a small child.'[16] Scofield suggests that the eagle also seems likely to be *Te Hokioi*, a massive black-and-white aerial predator with a red crest and yellow-green wingtips described to Sir George Grey, an early governor of New Zealand and later its 11th premier. A British soldier, explorer and writer, Grey was no ordinary colonial. He learnt the Māori language and was able to persuade Māori authorities to write down their legends and history. While no longer with us, Haast's eagle at least lives on through the stories passed down.

Death by serpent

Are you, like many others, ophidiophobic? Does the idea of a long, thin animal, covered in scales and with a pair of fangs and a forked tongue, cause you to break out into a sweat and run the other way? If you are scared of snakes then take some solace in the fact that it is a phobia that makes an awful lot of sense. As a biologist I am duty bound to tell you all the usual wonderful things about snakes – and doing so is easy. They really are incredible creatures, superbly adapted for a form of movement that is mesmerising. Seeing a large snake disappear vertically up a tree trunk or glide effortlessly and speedily across shifting sands gives a real insight into the reach and scope of evolution. The venoms of some snakes are an incredible mix of molecules that have medical applications including treating strokes and heart attacks. They are good-looking animals too, if you can bear to look at them. While some are pretty drab and nondescript, many snakes advertise their potential harm through bright colours and markings. Snakes have successfully colonised deserts, rainforests, woodlands and savannas and there are even around 70 species of snake that live in tropical seas. Yes, some snakes produce some truly horrific venom, and snake bite is a significant cause of death in some parts of the world, but most species are not dangerous and most of those that are tend to keep to themselves, with bites being a defensive measure. As is the case with the other animals I've covered so far, in reality we are far more dangerous to them than they are to us. However, this book focuses on predators that can eat us, and to eat a human being requires a very special snake. Or, more accurately, a very large one.

Defining what we mean by 'large' is complicated by the fact there are at least three measures we can use for snakes: length, girth and weight. The venomous snakes are not great candidates for hunting and eating us because they are generally too small. They simply lack the size to deal with a human even if the venom injected via biting proves fatal. That said, the longest venomous snake is usually accepted to be the king cobra at around 4m (13ft 1in), with some exceptional individuals achieving over 5m (16ft 5in). That is an impressive snake for sure, and with an untreated fatality rate of up to 60 per cent for human bite victims it can clearly take care of the killing side of predation.[17] Once we are dead, though, we won't be sliding down its throat. Prey-wise, king cobras mostly hunt other snakes, which are altogether slimmer meals than a human. Even the heavier Gaboon viper is nowhere near the right dimensions to make lunch of us. As is a common feature of 'big snakes', quoted maximum sizes of Gaboon vipers vary, but it is safe to say they are no longer than 2m (6ft 7in) (and most will be under 1.5m/4ft 3in), no heavier than around 10kg (22lb) and have a circumference of less than 40cm (1ft 4in). Make no mistake about it, these are impressive snakes and not a species to tangle with lightly. Gaboon viper venom can cause a troubling array of symptoms including pain, swelling, shock, blistering, hypertension, heart damage and severe tissue necrosis that can require surgical excision or amputation. Fatalities can occur, but at no point could anyone ever be eaten by a Gaboon viper, or indeed any other venomous snake.

The same cannot be said for the constrictors. These snakes, notably the boas and pythons, kill their prey by wrapping coils around it and, basically, squeezing hard. If you have ever handled even a small python you will be aware of the power that these animals can exert through their coils. Being squeezed in this way causes all manner of irreversible physical issues. The blood supply to the heart and brain can be shut off because the heart cannot pump harder than the constriction pressure being applied. Breathing is also affected, as is blood pressure. The prey quickly loses consciousness and thereafter the heart stops beating. The prey is then swallowed whole. To be able to squeeze a human to death, then swallow the body, requires a very large snake.

Of the large constrictors around the world there are four species that are worth some consideration as potential human-eaters: the green anaconda of South America; the African rock python of sub-Saharan

Africa; and the Burmese and reticulated pythons of South and South East Asia. Validated examples of humans being killed *and eaten* by any of these snakes are pretty thin on the ground but accounts of deaths caused by some of these species are easier to find. Burmese pythons are widely kept as pets and there are multiple stories of owners being killed by them. Grant Williams of New York City died in 1996, for example, when his 4m (13ft 1in) long Burmese python attacked him. But he wasn't eaten. When he was discovered by a neighbour he was certainly either very nearly or very actually dead (he was formally pronounced dead after being taken to a nearby hospital), but the snake was still coiled around his torso.[18] Hapless snake-keepers have also succumbed to African rock pythons. Dan Brandon was killed in 2017 by a 2.5m (8ft 2in) African rock python called Tiny that he kept in his home in Hampshire, UK. The first instance of a snake killing an owner in the UK, the incident was interpreted as 'affection' gone wrong during a handling session.[19] Whatever interpretation we choose to make of Tiny's behaviour, we can be in no doubt that such snakes are more than capable of killing humans given the opportunity to do so.

The African rock python is a potentially big snake. It can grow to more than 5.5m (18ft) and its diet includes larger animals such as warthogs and some antelope species. A 4.5m (14ft 9in) python killed and ate a juvenile waterbuck about 50m (164ft) from a bush tent in South Africa in which, most years, I spend a fortnight on a field course. A video of the incident was sent to me with the message 'that waterbuck looks about your size'.

It was probably a bit smaller, to be honest, but watching the lifeless body of a large antelope being slowly enveloped by a massive snake it felt churlish to argue over a few kilograms. However, despite being a credible size to eat us, verified reports of African rock pythons in the wild killing and then eating someone are exceptionally rare. Perhaps the best example we have is of a 13-year-old boy who was killed by a 4.5m (14ft 9in) python in the Waterberg District of Limpopo in South Africa.[20] The boy was killed, but the snake released his body when a man intervened with a pickaxe and some rocks. Although the victim wasn't swallowed, his head was covered with saliva, consistent with the snake considering the interaction as a feeding opportunity. The boy was 1.3m (4ft 3in) tall and weighed 45kg (99lb). This is smaller than the waterbuck taken by a similar-sized snake, which certainly points to a snake being able to take a small adult given the opportunity. Although the boy wasn't eaten, there seems no reason not to treat this attack as predation.[21]

Reticulated pythons prove more fertile ground for finding validated examples of humans falling prey to snakes. Native to South and South East Asia, these snakes can be even longer than African rock pythons, with several examples of individuals exceeding 6m (19ft 8in). Again, reports of people being killed are more common (although still rare) than reports of being killed and eaten, but for this species at least we can definitely become lunch. In 2017, a man called Akbar went missing while harvesting palm oil on the Indonesian island of Sulawesi. A search ensued during which a reticulated python reported to be 7m (23ft) was

found. Snakes are hard to measure by eye and rarely
have the manners to present themselves stretched out
by a measuring tape, at least when they are still alive.
There is also a natural human tendency to exaggerate,
so 7m (23ft) is a length that needs to be treated with
caution. However, such a length is lent some credibility
by a report of a reticulated python in East Kalimantan,
Indonesian Borneo.[22] This snake was reliably measured
at 6.95m (22ft 10in) after being captured unable to move
with a belly full of sun bear (the snake was later released
unharmed). Regardless of the exact dimensions of the
snake in Sulawesi, the fact that it had killed and eaten
Akbar is beyond dispute. The snake was cut open and
the man's body, fully clothed, was found inside. The
procedure was videoed and stills are presented in many
of the accounts of the incident, which was widely
reported by the world's media.[23]

There are other reports of reticulated pythons
potentially attacking and eating people, but perhaps the
most revealing report in terms of our relationships with
predators concerns a study of the Aeta Negrito people
of the Philippines. A collection of more than 7,600
islands in the Western Pacific, the Philippines is often
overlooked as a nation but was home to 109 million
people in 2020, making it the 12th most populous
country in the world. The Aeta people (also called the
Agta or Dumagat) are thought to be one of the earliest
inhabitants of the Philippines and historically were
nomadic hunter-gatherers.[24] By 1990 they had
transitioned to a settled peasant lifestyle, but were still
living as preliterate hunter-gathers when an
ethnographic study was undertaken in the 1960s.[25] An

aspect of this study examined the relationship between the Aeta people and reticulated pythons, in particular looking at attacks.

Staggeringly, 26 per cent of men reported being attacked by pythons in the forest, compared with just 1.6 per cent of women. Other results were no less surprising. The authors of the study found that, 'Giant serpents had attacked 14 Agta once each and two Agta men twice for a total of 18 non-fatal attacks. Fifteen (81.3 per cent) of the Agta attacked had sustained python bites; 11 exhibited substantial scars on the lower limbs or less frequently, hands and torso … Men generally were struck while walking in dense rainforest seeking game and useful plants, and they thwarted attacks by dispatching snakes with a large bolo knife or homemade shotgun.' A bolo knife is a large tool used like a machete, but with a blade that curves and widens at the tip. If you had to pick a knife with which to tackle a snake, you couldn't choose much better. Homemade shotguns, on the other hand, are a risky weapon for all concerned.

Earlier I suggested that attacks by large constrictors were rare, but the Philippine study shows us that, when living entirely within python habitat, such interactions may be far more common. Those attacks can also be fatal as the study showed: 'Nineteen (15.8 per cent) respondents had known at least one Agta killed by a python for a total of six specified fatalities.' At least one of these fatalities included consumption of the victim, whose body was later recovered from the snake's digestive system. Furthermore, 'Fifteen respondents recalled that a python entered a thatched dwelling at sunset (not yet dark), killed two of three sibling children, and was coiled around and swallowing one of them headfirst when the father returned and killed the snake with his bolo; the third child, a six-month-old girl, was uninjured.'

Injury or death by snake was a relatively common feature of life for the Aeta during their hunter-gather phase. Metal weapons allowed them to fight back, but that would not have been the case a few centuries ago before such weapons appeared. At that point Aeta deaths from python predation have been estimated to have perhaps exceeded 8 per cent. Clearly one factor in that vulnerability is the fact that they lived in forests alongside large pythons, which brings us back to the familiar theme developed in earlier chapters: it is the rural poor who bear the brunt of wildlife attacks. However, there is another factor at play with the Aeta: they are small. A study of their eyesight provides incidental data that gives us insight into their average height and weight.[26] Considering men, who are far more likely than women to be attacked, and taking middle-aged men as our focal group, then their average height is 1.49m (4ft 11in). Their average weight is 48.5kg (107lb). I am not an especially large man by modern standards, but I'm well over a foot taller and close to 35kg (77lb) heavier than an average Aeta man. Reticulated pythons have been reported to kill and eat 60kg (132lb) pigs, so while they might think twice about me (and this isn't something I'd want to test), an Aeta man is well within their natural prey range. Incidentally, early hominins such as *Homo erectus* and *Australopithecus africanus* would also have been comfortably within the prey spectrum of large constrictor snakes. Perhaps the Taung Child was too busy looking down for snakes to look up for eagles?

The last species of giant snake capable of killing and eating us is the green anaconda. The largest snake in South America, the anaconda is not as long as the reticulated python, very rarely exceeding 5m (16ft 5in), but has far more girth, making it a heavier snake. It spends much of its life in and around water and is a powerful swimmer. The anaconda has long been the source of legendary reports of truly gigantic individuals of 12m (39ft 4in) or more, but there is no reliable evidence to back up such claims. Nonetheless, they are very large snakes and, like the pythons I've already discussed, would be capable of killing and swallowing a human. However, unlike those python species there are no fully verified accounts of it happening.[27] They certainly could, and I'd be willing to bet they have, but we just don't know for sure. Of course, within the range of these snakes there are about 50 indigenous peoples living in isolation from the modern world, as well as some groups still defined as uncontacted people. As with so many fatal interactions with animals, those most likely to become victims are among the least likely to have their fate recorded.

CHAPTER SIX

Hyenas

The large predators I've discussed so far are reasonably familiar to most people. While depth and breadth of knowledge may vary, there are many features of lions, tigers and crocodiles that are familiar enough to treat them as almost universally known facts. Lions live in groups called prides, tigers live in forests in India, crocodiles live in rivers in tropical countries, saltwater crocodiles are massive, tigers are stripy and so on. The next group of animals I'm going to consider is certainly familiar to most people by name, but further familiarity is likely constrained to crude caricature. Worse still, that caricature is likely filled in with some colour provided by their villainous role in Disney's *The Lion King*. The word 'hyena' may be very recognisable, but the animals themselves are much less well known. Massively misunderstood, erroneously maligned and in general treated as craven, worthless scavengers, the bad press following these animals is hardly helped by the fact they have a complex relationship with us that, unfortunately, can involve predatory attacks.[1]

Before examining our tricky relationship with hyenas, it is important to gain an appreciation of the animals themselves because they are, in many and surprising ways, rather unusual. There are four extant species of hyena, but their fossil record indicates a far greater diversity over the past 20 million years or so. It is worth a small diversion into their evolutionary history because it reveals why, despite being superficially similar in appearance and ecology to canids ('dogs' – see Chapter Nine), hyenas have very different ancestry.

Hyenas evolved from small arboreal (tree-living) mammals that would have been similar to the modern-day African civet. The African civet looks a little like a cross between a raccoon and a domestic cat, with a striped tail and a pale coat covered in dark blotches. They are 'rear-wheel drive' animals, with disproportionately large hindquarters, and they possess a band of erectile hairs down the back that can be raised to form a crest. This crest makes the civet look larger and fiercer, and can be raised when the animal feels threatened. The civets (there are a number of different species that can be found throughout African and Asia) belong to a wider grouping of carnivores called the Feliformia. This group also includes the mongooses, the fossa of Madagascar (worth a quick Google if you've never seen one), the genets (worth a Google for the cuteness factor) and the cats, including the big cats. So, although quite dog-like, hyenas are actually more closely related to cats, but are much more closely related to civets.

An early hyena genus, *Plioviverrops*, lived 20 million years ago and was very similar to, but had already diverged from, the civets in some of the key areas (such as dentition and the structure of the middle ear) that distinguish the hyenas. The lineage did well, getting larger and more diverse until eventually it split into two distinct lines, known as the bone-crushing hyenas and the dog-like hyenas. The dog-like hyenas thrived for a while, but by around 1.5 million years ago they had mostly gone extinct, perhaps because they were unable to compete with the various species of canid that had evolved by then. The only species of dog-like hyena left now is the enigmatic aardwolf. Sometimes called the civet-hyena (and there is a definite resemblance to the African civet), the aardwolf gets its common name from the Afrikaans for 'earth-wolf'. These relatively small animals – they are less than 1m (3ft 3in) long and weigh around 10kg (22lb) – eat insects, mainly focusing on termites. They have a long, sticky tongue that helps in this task and they are reported to consume up to 250,000 insects in a night. Aardwolves are secretive, nocturnal creatures and can be tricky to see even when they are known to be present in an area. Aardwolves need not concern us further because they are absolutely no threat to us. They are small, timid and, crucially, don't eat meat in any sense that should be worrying unless you are a termite.

The remaining three species of hyena belong to the bone crushers – solidly built with broad, strong skulls and large teeth resembling those of a dog. The spotted hyena (the species caricatured by Disney) and the brown hyena are confined to sub-Saharan Africa, while the smaller striped hyena is also found in northern and northeastern Africa as well as the Middle East, Central Asia and the Indian subcontinent. Although the striped hyena more closely resembles the aardwolf than it does the spotted and brown hyenas, it is longer (by as much as 50cm/1ft 8in) and much more heavily built than its termite-eating cousin, weighing around 35kg (77lb) or more. Predominately a scavenger, the striped hyena will take live prey on occasion, although overall its diet is perhaps best described as 'whatever it finds'. It will eat grasshoppers, fruit, tortoises, porcupines, carrion, rubbish and pretty much anything else. Hyenas have powerful digestive secretions that allow them to consume food that is long past its best-before date.

Striped hyenas will supplement their scavenging by hunting live prey – and when they do it isn't pretty. Too small to overpower most large prey animals, striped hyenas instead run their prey down until they can bring their teeth to bear on some soft tissue around the groin or under the abdomen. Once they get a hold on a tired animal it is the beginning of a sometimes very protracted end, as the predator works on eviscerating and tearing the animal apart, often eating portions while the prey is still alive. This form of predation is similar to the way that African wild dogs and free-ranging domestic dogs in India kill their prey (see Chapter Nine). It is the brutal reality of real-world predation, where most animals aren't killed neatly. The TV documentaries showcasing lions delivering an elegant and rapid *coup de grâce* on a small gazelle are heavily sanitising real predation. I well remember seeing a wildebeest limping across the veldt in South Africa, dragging much of its gut through the dirt below it. It had fallen victim to a predatory attack, with the predators (very possibly spotted hyenas) waiting until it died, or at least until the kick had gone out of it, before resuming their feast.

Larger than the striped hyena, and more sombrely marked, is the brown hyena. I have a soft spot for brown hyenas. Very much a southern Africa species, they were the first hyena I saw in the wild and for some reason I seem to be blessed with remarkable good fortune when it comes to seeing this relatively uncommon species. In a memorable sighting I watched a denning brown hyena cautiously bring her cubs out into the early evening light, and the sight would have turned even the most iron-willed hyena-hater towards the light. Brown hyenas have a shaggy coat that is, as you might expect, dark brown all over, except for cream-coloured stripes running around the legs and a lighter, cream-brown ruff around the neck. They are larger and more heavy set than striped hyenas, weighing in at around 40kg (88lb). Opinions vary, but to my mind they are the most beautiful of all the hyena species.

Likewise, the spotted hyena is probably the least beautiful, but most impressive, hyena. The quintessential 'laughing hyena', it is the largest species, sometimes exceeding 60kg (132lb). If the African civet is a 'rear-wheel drive' animal then the spotted hyena is very much 'front-wheel drive'. Its shoulders and neck are well-developed and powerful, while the hindquarters are much less impressive. This smaller back end gives the animal a heavy-set, raked-back look. Covered in a short coat, the dark spots that give the species its common name stand out in individually identifying patterns against a background of rather nondescript, greyish-yellow-brown fur. It is, though, neither the colour of the spotted hyena nor the patterns of its fur that draw the eye when you see one. The inevitable point of focus is its neck and head. That head is bulky. Broad and deep, the large size is a consequence of the heavy skull and the huge musculature it supports. There is a pronounced ridge of bone running from front to back (the sagittal crest) to which attach the powerful temporalis muscles involved in biting. The head is all about bite force, with muscles contracting to bring together huge conical molars (the bone crushers), formidable canines (for stabbing and tearing) and sharp, flesh-cutting carnassials. The entire skull is strengthened and vaulted to prevent it from buckling under the forces generated by the muscles.[2] Hyenas can easily crush and break bones, even the larger, heavier limb bones of animals like giraffes. A large head, of course, needs a big neck and powerful shoulders to support it and to help with all the ripping and tearing. Hyenas really are well put-together animals and I've yet to meet a biologist who isn't a fan. If you haven't guessed it yet, I'm more than a little enamoured with them. And that's even before considering perhaps the most unusual feature of their biology; the female's pseudo-penis.

The simplest way to put this is that female spotted hyenas appear to have a penis.[3] And a decent-sized penis at that. In fact, the female's pseudo-penis (as it is properly termed) is almost as long as the male's actual penis, with a similar girth, about 17cm (6.7in) long and just over 2cm (0.8in) in diameter. Just like the male's penis, the pseudo-penis is erectile, and that gives us a clue as to its anatomy and origin. The pseudo-penis is a very much enlarged clitoris, with well-developed erectile tissue. Behind the clitoris is what appears to be a scrotum (a 'pseudo-scrotum') formed by fusion of the labia. I am afraid I have to take someone else's word for this, but apparently the most reliable way to determine the sex of a spotted hyena is to palpate the scrotum to determine the presence of testicles (male) as opposed to fatty tissue deposits (female).[4] The vagina runs through the enlarged clitoris, so females give birth through their pseudo-penises. The position of the vagina also means that it is more or less impossible for a male to force copulation on a female, a task made more difficult still by the fact that spotted hyena females are often larger than males and dominant to them. Unlike the other hyena species, spotted hyenas live in large – sometimes very large – social groups. Brown hyenas have 'clans' of perhaps four to six individuals, but spotted hyena groups can have more than 50 individuals associated with them, and there are records of groups of 130 or more.[5]

More predator than scavenger

Spotted hyenas are excellent hunters. Due to their body size and the fact that they form large groups, they are able to take down large prey. Although best known for scavenging, it is estimated that in some regions more than 90 per cent of some species' diet is meat they have killed themselves. Their predation technique for larger animals is the disembowelling method described earlier, but for small- and medium-sized prey (and we fall into that latter category) they go for a different approach, attacking the head and neck regions. Scavenging is still a potentially important source of food though, and spotted hyenas play an ecologically vital role at carcasses, breaking them open and cracking bones in ways that other scavengers like vultures can't. This role aside – and despite what Disney would have us believe – unlike any other hyena species, spotted hyenas are best thought of as predators.[6] As a consequence, it is the spotted hyena that is the species most commonly implicated in human attacks.

It would be inaccurate to portray the spotted hyena as a major threat to humans, especially when we contrast it with the big hitters in this regard, the lion, tiger and crocodile. However, predatory attacks, although uncommon, are not abnormal. In common with other predatory species, though, it is their role as livestock predators that most often places them in conflict with people. There is no doubt that spotted hyenas can be very effective killers of livestock and domestic animals. One study of hyenas in the Tigray region of northern Ethiopia interviewed 1,686 households, asking them about their experiences with the species between 2005 and 2009.[7] The study found that 492 domestic animals were taken during that period. This included cats, dogs and poultry, but when excluding these smaller species the total loss came to around 1 per cent of available larger livestock in the area. That might not seem much to you, but bear in mind that the 1 per cent losses across a population of livestock may not be felt evenly by those who own the livestock. If you only own a few cattle and you lose three to hyenas then you face potential economic ruin and genuine existential threat. It is also worth bearing in mind that these losses occur despite measures being take to prevent them, such as enclosing livestock at night in a *kraal* or *boma* (of varying build quality) and actively guarding livestock against predators.

As we have already seen, loss of livestock is often a major driver of human–wildlife conflict and while some losses can be tolerated there is inevitably a threshold at which that tolerance wears thin, and then disappears. For some species, in some areas, that threshold is high. As we saw with crocodiles in Chapter Four, our cultural relationships with predators can play a large part in mediating potential conflict and lead to outcomes that might include a very high degree of toleration even for species that attack and consume people. The pattern of human–predator coexistence across the world is nothing if not complex, made more so by the diversity of predators, the exact form our interaction with them takes, and our social, historical, financial and cultural relationships with them. These are general themes we will return to in the final chapter – and of course visit again as we explore other predator species.

Benefits and costs

Hyenas enjoy a reasonable degree of human toleration in parts of their range despite the fact that all three 'bone-crusher' species are recorded killing livestock and that spotted (and to a far lesser extent striped) hyenas represent a physical threat to people. The key to this tolerance is the very behaviour, scavenging, that makes many people squeamish about hyenas. Spotted hyenas will feed on virtually anything, which means that rubbish, dung, carrion, dead livestock and other human-produced waste can be cleared up very effectively. In fact, to hyenas, humans represent an excellent source of potential food resources just from the components of our lives that we discard. In Ethiopia – and in other parts of their range – spotted hyenas have become to some extent a 'peri-urban' or even urban species in much the same way that foxes and gulls have moved into our towns and cities. We will see a similar movement into human settlements in the next chapter when we consider leopards.

The simple removal of waste is one tangible benefit of having scavengers around, but if that waste is somehow harmful, perhaps by being a source of disease, then its removal can have public health benefits. A recent study of hyenas examined the potential public health benefits of the presence of hyenas in urban areas that were actively scavenging livestock carcasses.[8] By eating an estimated 207 tonnes of carcasses annually, some of which would be diseased with anthrax and bovine tuberculosis (bTB), hyenas in this study are estimated to have saved five anthrax and bTB infections in people, and 140 infections in cattle, sheep and goats. The authors of the study were able to put a financial value on the health benefits of hyena presence based on the treatment costs and livestock losses avoided: US$52,165. The authors also recognise the potential of hyenas scavenging livestock to prevent the 'spillovers of novel zoonotic diseases into the human reservoir', which is an observation made more pertinent by the Covid-19 pandemic.[9]

There may be other, less obvious benefits to having hyenas around. In an account of ongoing interactions between people and hyenas in the East Hararge region of Ethiopia, anthropologist Marcus Baynes-Rock recounts a number of attacks, especially on children.[10] At least one of these attacks was fatal, with the victim, a young girl, being 'eaten down to the chest'. The picture he paints is one of ongoing conflict between humans and hyenas, with each side inflicting fatalities on the other. He tells of a doctor treating 30 cases of hyena attacks in 14 months. These are not 'abnormal' events and – while individually the chances of being attacked may be relatively small – they are far from insignificant. Thirty per cent of the people Baynes-Rock spoke to across eight villages personally knew someone who had been attacked. Attacks on people spur on retaliatory killing of hyenas. Killings might involve snaring or poisoning, but more commonly involve a large group of people chasing, exhausting and eventually cornering a hyena, which is then set upon with axes and other tools. However, this is not a lawless Wild West of animal killing. Just randomly killing hyenas is illegal. If a farmer kills a hyena then it is only legal if the farmer can prove a direct threat to life or property and that the killing is a response to that threat.

Despite local residents knowing where the hyenas den, and having the ability to kill them, hyenas continue to persist in the East Hararge region. This is doubly surprising, since they were considered vermin up until 1998 and afforded no protection at all. As with the Scotts' study of crocodiles in Tanzania (see Chapter Four), Baynes-Rock asks why predators continue to coexist with people despite the relatively high human cost of their presence. He acknowledges the possible benefits of their scavenging on human waste and diseased livestock carcasses as one reason, but concludes that a greater reason that hyenas are tolerated is because of the crop *Catha edulis*, also known as khat, chat or qat. The leaves of this plant are chewed and this releases cathinone, a substance that provides the chewer with a sense of euphoria, loss of appetite and a general stimulating effect similar to amphetamine. Khat is widely used throughout East Africa and beyond, and is grown extensively in the region. Hyenas are seen as a deterrent against theft (khat is a profitable crop) and what is more they also deter dik-diks and other herbivores from eating the crop. Indeed, hyena scat is powdered and sprayed on to crops to deter these herbivores, a fact worth remembering if you ever decide to chew on khat. The people of East Hararge have another reason to tolerate hyenas – they are believed to eat evil spirits. When Baynes-Rock asked 140 people if hyenas were a benefit to the area, 88 per cent either agreed or strongly agreed that they were.

Of course, the benefits of having carnivores living close to humans are countered by any costs and the balance point of that cost–benefit seesaw depends very much on the impacts either felt, or perceived to be present, by local communities. Such impacts and their effect on communities are likely to be as varied and diverse as the communities themselves. Returning to the study of human–hyena interactions in Tigray we find that potential disease prevention is not a feature of hyena presence that comes high on the agenda of the villagers questioned. Even against a background of relatively very small livestock losses, farmers in that region would still rather hyenas weren't present; 70 per cent of those surveyed did not want to conserve spotted hyenas. One reason for this might well be the threat they pose to people. During the survey, ten attacks on people were reported, nine of which occurred at night. As with other predatory species we've encountered there was a pronounced sex difference, with nine of the victims being male. Half of the attacks were on people sleeping outside, while other victims were defecating outside, helping another person being attacked or, in an echo of the lions of Tanzania, the hyenas entered dwellings to attack. None of these attacks was fatal and therefore none resulted in predation, but the inevitable conclusion is that the attacks were likely to be predation-motivated.

The authors of the Tigray paper don't provide any follow-up information of the reported attacks, but their study does clearly show some consumption of humans by hyenas – although not in the way we might think. The authors looked directly at what hyenas ate by examining scats. By identifying what species hairs in the scat had come from they were able to infer what hyenas ate. What they found was interesting in at least two respects. Firstly, none of the 1,200 hyena scats they examined contained any hairs from wildlife species. There were, however, plenty of mammal hairs present. Most common were sheep, followed by horse, donkey, cattle and then goat. Some would have come from scavenged carcasses, but many would come from the livestock reported as being predated by the survey respondents. The absence of wildlife hairs is largely explained by the absence of wildlife in the area. This, sadly, is the pattern across many other parts of Ethiopia where hyenas are present, so the pattern of livestock predation found in Tigray probably applies in general elsewhere. The second thing they found, in 5.5 per cent of the scat analyses they undertook, was human hair.

The presence of human hair in scat samples can only mean one thing: hyenas are eating people. However, the route by which that hair entered the hyenas' notoriously robust digestive system is far less definitive. During the course of the study there were no reports of anyone being killed by hyenas, but this doesn't rule out predation. It is perfectly possible that someone may have fallen victim and gone unreported in the study. However, the 10 reports of attacks and the large number of people interviewed suggest that this is unlikely. A better explanation in this case is that the hair was ingested through scavenging in cemeteries and rubbish dumps. Ingestion of humans, but not predation.

While the hyenas of Tigray may not have killed and eaten anyone, the 10 attacks reported during the study period certainly point towards that possibility. Attacks on humans by hyenas often follow a similar pattern, with the initial attack focused on the face, head and neck regions. This, of course, is the same strategy they use when taking down medium-sized prey. Bristol-based facial-reconstruction surgeon Matt Fell led a team that studied people who had fallen victim to hyena attacks in rural eastern Ethiopia.[11] The study examined four victims of hyena attacks – three male and one female. At the time of their attack, the males were young (five, seven and 15) and the female relatively old (50). All were attacked as the sun was setting or at night. Each victim had a particular vulnerability. The five-year-old boy was the smallest child in his group; the seven-year-old was playing alone; the 15-year-old was sleeping alone while guarding cattle; and the 50-year-old woman was the slowest runner in the group she was in. All four were attacked by a solitary hyena.

All four victims suffered from considerable, disfiguring facial injuries. The paper contains photographs and descriptions of the injuries, which resulted from the attacking hyena closing its powerful jaws around the softer structures of the face, across the skull and around the scalp. There is often soft tissue loss as well as injury, degloving of the scalp (*i.e.* removal of the skin so that it is hanging off and exposing the skull beneath) and considerable damage to the mouth, nose and cheek regions. There may also be fracturing of the underlying bones in the facial region. In each incident, the hyena seems to have attacked with open jaws, lunging at the victim, closing its jaws on the facial region and tearing away flesh in much the same way that it would tackle other medium-sized prey.

The people in the study were lucky in some respects. They were being treated by Project Harrar, a community-based charity that coordinates surgical treatments in Ethiopia. Many attack victims may be far less lucky. The injuries themselves are bad enough, resulting in a mix of facial scarring, excessive eye-watering, pain, and difficulty chewing and breathing, but the facial scarring carries further problems. All four victims reported issues arising from facial scarring that manifest as verbal insults, subdued mood, abstinence from school (as a consequence of verbal insults from peers) and other social issues. The woman who was attacked was unable to find employment because of her facial scarring. Thus, even non-fatal attacks can have serious life-changing and life-lasting consequences.

The conclusion of the authors of the facial injuries study in Ethiopia was that these attacks were predatory in nature. I think it is hard to disagree. The injuries strongly suggest predation and the hyenas were not under any threat; these were not defensive attacks. Hyenas hunt cautiously, selecting weaker individuals by appearance or behaviour. This pattern fits with the attack profiles of the victims. Furthermore, hyenas tend to take prey in low light or at night, which fits with the timing of the attacks. Taken overall, there is no reason to suppose that the hyenas involved in these attacks were not acting as predators and it seems clear that the victims were their intended prey.

Other hyena attacks are well documented and, though rare (as is the case with all the species covered in this book), they are by no means insignificant, especially to those living alongside them. The urban hyenas of Addis Ababa, Ethiopia's capital, are well known to attack rough sleepers, sometimes gnawing on the fingers and toes of people incapacitated by drink or drugs. One report tells of a man brought to a clinic with much of his scalp hanging off after an attack, echoing the reports from those attacked in more rural areas. Given these reports it seems highly unlikely that hyenas in the city aren't killing and eating at least some homeless people, whose deaths will go unrecorded or perhaps be put down to scavenging. The pattern of targeting the vulnerable is once again evident; in 2013, the BBC reported a that a mother camping outside a church near the Hilton Hotel in Addis Ababa had her baby taken and killed by a hyena.[12]

Given the way that hyenas hunt, lacking the technical skill to dispatch prey quickly as tigers and lions can, it is not surprising that human victims of attacks may survive. People are able to get help, to be treated and to summon others to ward off further attacks. Thus, while hyena attacks can result in very major injuries, fatalities are rarer. As with other predators, attacks often involve children. Musa Jelle, a 10-year-old boy, and his family were attacked by a pack of hyenas in their home in the town of Wajir in northern Kenya. Two children were killed in the attack, but Musa survived, albeit with major facial scarring.[13] Nine-year-old Rodwell Khomazana, from Zimbabwe, become the focus of global media interest in 2021 following a serious hyena attack in May of that year.[14] His facial injuries were severe, but Khomazana was able to be treated by top surgeons in South Africa, who fitted him with a prosthetic nose and eye.[15] Six-months prior to his attack, a fellow Zimbabwean was less lucky, showing that attacks on adults can occur and can be fatal. In November 2020, 46-year-old Tendai Maseka was attacked by a pack of spotted hyenas in his hut in the village of Bangure. His arms, legs and head were found the next day, but the rest of Tendai had been consumed. He was alone, and (with a nod back to Chapter Two) had last been seen drinking. Tracks in his room suggested a struggle but, since he was heavily outnumbered, there could only have been one outcome.[16]

Moses Lekalau, a Kenyan herdsman, also fell victim to hyenas, but in a particularly unusual incident that underlines the opportunistic nature of hyena attacks. Moses was walking back to his homestead with some cattle along a forest path in Maralal, 257km (160 miles) north of Nairobi. Maralal is a popular tourist destination, with visitors mostly coming to see the Maralal National

Sanctuary, which houses a wide variety of Kenyan wildlife species. Outside the sanctuary, other species roam. On his walk home, Moses was attacked by a lion. It would seem that Moses put up quite a fight. He eventually managed to spear the lion and then clubbed it to death in a struggle that was reported to have lasted half an hour. As he recovered from this initial attack, he was found by a group of hyenas. They presumably sensed his weakness and launched their own attack. A passing motorist managed to chase the hyenas away and then drive a badly wounded (and frankly incredibly unlucky) Moses to hospital. Moses underwent surgery that involved amputating his arms and reconstructing his face. In a now familiar pattern, Moses's face had been partially eaten by the hyenas, which also accounted for most of his other injuries. Despite his experiences, Moses was able to talk on arrival at the hospital, but he died shortly after his surgery from blood loss.[17]

A flexible and versatile species

Hyenas are superb opportunists and in some cases – as the hyenas of Addis Ababa demonstrate – are more than able to adapt to a human-dominated landscape. Their ability to scavenge and hunt gives them a tremendous flexibility not enjoyed by other predators. When humans deplete natural prey indirectly through conversion of habitat – and directly by killing animals for food – lions and tigers must turn to livestock, and sometimes to humans, to feed themselves. This prey switch has inevitable negative consequences for all concerned. By having a more mixed diet, hyenas have more options.

One very unusual study that underlines how flexible hyenas are when it comes to what they eat examined how human religious activities influenced their diet. In northern Ethiopia many people are members of the Ethiopian Orthodox Tewahedo Church. The keeping of fast days (and runs of fast days) is an important tradition and there are a large number of fasts scattered through the year. The longest fasting period is a 55-day run that precedes Easter and is equivalent to Lent. *Abye Tsome*, or *Hudade* as it is also known, calls for a cessation of meat eating. What this means is that across regions observing the fast there is a sharp, near-total decline in demand for meat. This in turn leads to a decline in the availability of slaughtering waste, which is an important component of hyena diet in this region.[18]

Before the fasting period, analysing hyena scat showed that cattle and sheep hair were the most common, supporting the idea that hyenas are feeding on slaughterhouse waste, although also possibly supplementing with some low-level predatory activity. However, during the fasting period the profile of hairs found in scat changed dramatically, with donkey hair becoming most common. When slaughter waste was not available, hyenas rapidly switched to a prey item that was. Donkeys are plentiful in the region, and unlike cattle and other livestock they are usually kept outside any protected compound at night. Donkeys, which tend to suffer the worst of human 'kindness' in many parts of the world, may also be abandoned when they are too weak to be useful work animals. They make ideal prey for the opportunistic spotted hyena. After the fasting period, normal service is rapidly resumed, with cattle once again taking the top spot in terms of hairs in scat.

Opportunism and flexibility are key to hyena success, but also mean that we, comfortably within their prey spectrum size-wise, can feature in their diet both as prey and as carrion. As the killing and eating of Tendai Maseka shows, spotted hyenas' tendency to be in large social groups can also make it easier for them to predate us if they decide to take the opportunity. The study of attack victims in Ethiopia, where all four people were attacked by a single hyena, demonstrates that hyenas don't need strength of numbers to attack, but such attacks are less likely to be fatal.

Another factor that may be involved in determining attacks is hyena size. Perhaps the most notorious incident of hyena predation of humans occurred in the late 1950s in the south of Malawi, close to the Mozambique border. The population of the region at the time was around 10,000 spread across an area of about 2,000 km² (772 square miles). Every dry season, when due to high night-time temperatures people slept outside on verandas, hyenas would grab victims and drag them off into the bush. Many of the victims were children. In total, 27 people were killed and eaten by hyenas in the region across five years. In 1962 a game warden called Mr Belestra reported shooting two hyenas after attacks in the town of Mlanje. He describes some of the attacks and what remained of the victims, and there can be no doubt that these were predatory attacks.[19] Such a spate of attacks is extremely unusual and, in this regard, it is notable that both hyenas Belestra shot were very large. A typical adult spotted hyena weighs somewhere between 45 and 65kg (99 and 143lb), but the two Malawi hyenas weighed 72kg (159lb) and 77kg (170lb).[20] Possibly being larger makes human predation that little bit easier, although I would suggest that even a small hyena is more than a match for most people.

The other hyenas
So far, I have focused on the spotted hyena, but there are two other bone-crushing species that could perhaps be capable of attacking, killing and eating humans. We can quickly dismiss the brown hyena as a potential human-eater, or even an attacker of people. Brown hyenas are smaller than spotted hyenas, live in much smaller groups and, crucially, are primarily scavengers. Brown hyenas are not hunters and although they will take live prey they do so rarely, and typically prey makes up less than 5 per cent of their diet.[21] The animals they do hunt are small, including bat-eared foxes and springhares. Brown hyenas are also timid and avoid people. Of course, they could in principle attack us and I suppose if an animal was especially hungry it might try, but well-documented cases of it happening are elusive. A comprehensive review of their biology and relationships with humans published in 1976 makes no mention of human attacks up to that point – and it doesn't seem that things have changed. Brown hyenas pose no threat to us.[22]

Striped hyenas, on the other hand, are a threat – albeit less of a threat than spotted hyenas. Their range is the most geographically extensive of any hyena species and it is the east of that range, specifically India, that provides a recent, and graphic, example of a striped hyena attacking people. In September 2021, a video emerged of a striped hyena attacking and injuring a 70-year-old man, Pandurang Sahadu Jadhav, in the Pune district of western India.[23] A 25-year-old man, Rahul Madhukar Gade, was also attacked and injured by the same hyena just 500m (1,640ft) from the location where Jadhav was attacked. But were these attacks predatory? It is hard to say for sure. The hyena was observed by tourists just before the attack on Jadhav and they described it as 'possibly unstable'. The tourists actually warned Jadhav not to go down the road, but he brushed aside their warnings because, as he told them, hyenas are common in the area. That Jadhav was happy to walk down the road in the presence of hyenas tells us much about the danger they are perceived to pose. The hyena that attacked Jadhav and Gade was reported to be injured (Corbett's injury hypothesis rearing its head again), and was probably hungry and dehydrated. The press reports from India were at pains to point out that such attacks are incredibly rare.

There are other reports of striped hyenas attacking and sometimes killing people, especially children. However, such reports are thin on the ground, at least in modern times, when the species is more restricted in range and abundance than it once was. A number of credible reports of human predation are available for parts of the Russian empire in times when the species was much more common. In the 1880s, a three-year spate of attacks was reported in the Erivan province in the Caucasus (now part of Armenia), with one year seeing 25 children wounded. The victims were mostly sleeping in the open. Fatal attacks in the region were also reported in 1908. The taking of children was still being reported into Soviet times in the 1930s and 40s, but such reports, though tragic, are unusual, spaced out through time and have more or less dried up as the species has become increasingly uncommon.[24] Striped hyenas can certainly kill us in predatory attacks, and undoubtedly have done, but they are not a real threat to us in modern times.

Hyenas are biologically fascinating animals. Spotted hyenas are relatively well studied, but our knowledge of brown and striped hyenas lags far behind, and of the aardwolf even further. Hyenas also suffer very badly when it comes to public image. Their depiction in *The Lion King*, as craven animals on the side of evil, is not the cause of their poor image but rather a symptom, a sign of thousands of years of cultural mistrust and suspicion. Folklore, myth and mysticism, along with superstition and fear, weave a complex web of inter-relationships between us and them that is especially the case for spotted and striped hyenas. There is a pervasive perception of them across their range as sly and cowardly, as grave-robbing ugly gluttons with a large dose of stupidity thrown in. Their habits of scavenging, including on our own dead, connects them with a number of beliefs relating to 'were-hyenas', vampiric behaviour and witchcraft. However, it is as scavengers and powerful carcass breakers that brown and striped hyenas play a crucial role in nutrient cycling. Spotted hyenas play two ecological roles, adding predation into the mix. If we prefer to view the world anthropocentrically rather than ecologically, then hyenas do an excellent job of clearing up our mess, and what is more may be helping to reduce disease at the same time. As we have seen, these 'services', along with crop protection and a belief in spirit-eating, can result in hyenas being tolerated and seen as beneficial in some parts of their range, outweighing the costs of having them around. But overall, despite islands of tolerance and understanding, hyenas largely remain victims of persecution and habitat loss. You would think that such important, interesting, useful and charismatic creatures would be as celebrated as lions and tigers. That they aren't says much about our attitude towards the natural world and the amount of growing up we will need to do if we are to prevent hyenas – and similarly 'ugly' and 'unpopular' species – from declining yet further.

CHAPTER SEVEN

Other Cats

A large cat is well set up for making us prey, as lions and tigers amply demonstrated in Chapters Two and Three. Their size, power, dentition and claws, together with their hunting ecology, combine to produce a very effective predator package. While lions and tigers are the biggest – and best known – of the large cats, there are similar species that can also be problematic for people.

The term 'big cat' is widely used to describe, well, big cats, but from an evolutionary perspective the name can be a little tricky. All members of the genus *Panthera* are definitely in the 'big cats' club, which means that lions and tigers are joined by leopards, closely related to lions, and jaguars, which are closely related to lions and leopards but more distantly related to tigers. These four species have always been grouped together, but recently snow leopards were also added to the genus, joining as a sister species to tigers. The snow leopard is a source of conflict with humans because it will take livestock, but there are no credible reports of wild snow leopards attacking or eating people. The same cannot be said for 'regular' leopards and, albeit to a lesser extent, jaguars. We will return to these two species later.

The other cat species usually included in the 'big cats' are not in the genus *Panthera*. The first of them, the cheetah, is a very familiar cat to most. A distinctive athletic build supports its title as the fastest land animal (if we disregard a stooping peregrine falcon), while a spotted coat and prominent black tear-like streaks down its face make the cheetah unmistakable from most angles. Cheetahs were once well distributed throughout Africa, Asia and parts of Europe, but are now found in eastern and southern Africa, central Iran and possibly a few other locations in Asia that remain unconfirmed. The cheetah is not aggressive towards humans and is relatively easily tamed. There is a rich history of tamed cheetahs being used by humans as a hunting animal, including depictions in Ancient Egyptian and pre-Islamic art. Cheetahs in the modern world can come into conflict with humans because of livestock depredation, but as is the case with most of the species in this book their story is generally one of decline and threat. Our old enemies – habitat loss, persecution and prey-base depletion – rear their heads yet again and as a consequence the cheetah is at the centre of a number of initiatives to increase its number and range. Captive cheetahs have been known to attack people, but there are no fatalities reported. I know at least one person who received a relatively serious injury from a cheetah while working with them, but they lived to tell the tale and there is no question of attacks having a predatory motivation. Like snow leopards, cheetahs just don't seem to view us as prey.

Another cat species that is sometimes included in the 'big cats', but that need not concern us much in terms of human predation is the clouded leopard of South East Asia. A relatively small species, the clouded leopard is beautifully marked with blotches and patches vaguely resembling clouds. It is this spectacular coat that is the cause of some of the species' woes, with hunting and trade in live animals and skins being an ongoing problem. Habitat loss and forest fragmentation, though, especially through commercial deforestation, are playing the biggest role in the decline of this species.

As is the case with cheetahs, there are reports of captive clouded leopard attacks on people. On 25 July 2002 in Tacoma, Washington in the United States, a four-year-old male clouded leopard is reported to have 'jumped on and scratched both arms of a handler at the Point Defiance Zoo during a behind-the-scenes tour ... A small group of children was nearby. The woman had to call for help to get the 43-pound [19.5kg] cat under control and her wounds were treated at a hospital. The same animal had scratched another handler's leg 18 days previously.'[1] On 30 October 1999 in New York state, a keeper at the Buffalo Zoo was 'pounced on, bitten, and clawed by a clouded leopard when she entered the 10-foot square cage for cleaning. The incident happened in front of about 30 visitors. She was treated at a hospital for leg wounds.' These attacks were not fatal, and seem highly unlikely to have been motivated by predation. The fact that clouded leopards and cheetahs have been known, rarely, to attack humans does lend some weight to the idea that they *might* hunt us, but these captive animal attacks are not representative of normal, wild behaviour. There is a report, though, from the Saptari district of Nepal in 2019 that seems to show an attack by a wild clouded leopard on five people at once, all of whom were injured in the attack. The newspaper report is short and light on detail. It is also illustrated with a photograph of a leopard rather than a clouded leopard, but illustrating a nature story with the wrong species is a common problem. So much so, that it has its own Twitter hashtag, #taxonomyfail. The fact that people were attacked is not in question, but we cannot piece together enough from the reports to determine whether this was a predatory attack or to be 100 per cent confident that the species was a clouded leopard. Interestingly, a footnote reports that, 'The injured leopard was taken under control by rangers from Koshi Tappu Wildlife Reserve and local police who have taken it to the wildlife reserve for treatment.'[2] Whether the animal was injured before or during the attack is unclear, but my money would be on this being an animal injured before the attacks took place.

The final non-*Panthera* big cat is a species that has more of a track record of attacking, and sometimes eating, people. The cougar, also known as the mountain lion, the puma, the catamount and a host of other names, is native to the American continent and at one time was widespread across much of North America, Mexico, Central and South America. It has the greatest geographical range of any large animal on the American continent and can be found high in the northwestern Rockies all the way down the tip of Argentina.

The American lion
In May 2018, two mountain bikers were riding on a backwoods trail in a remote area near North Bend in Washington state, around 50km (31 miles) east of the city of Seattle. A cougar started to follow them. At this point the bikers, according to the county sheriff's department, 'did everything right'.[3] A natural instinct would be to tear off on the bikes, but running away from a predator is

nearly always a bad idea. A running animal is a prey animal. It is much better to, as the Washington State Department of Fish and Wildlife put it on their website, 'convince the cougar that you are not prey but a potential danger'. The two riders dismounted, made a load of noise and did everything they could to scare the animal away. One even hit the cat with his bike after the cat charged. Initially, the approach seems to have worked and the cougar ran off. But, when the men got back on their bikes, the cougar attacked again. One of the men was bitten on the head and shaken. Cougars typically attack prey by going for the head and neck, driving their large canines into the spaces between the vertebrae, adding some shaking to ensure a quick kill. No predator wants their prey kicking around and causing injuries. The man who was attacked first was lucky to survive and he probably owes his survival to his less lucky companion. The second man ran, and the cougar dropped its first victim to give chase. The chased man was killed and his body was dragged some distance back to what was reported as being the cougar's 'den'. The survivor managed to ride far enough to get a phone signal and was able to get medical help. It is common practice to hunt cougars using dogs, which can follow scent trails to find them and then chase the cat into a tree. This is what happened in this case and the cougar was shot.

There are many other reports of cougars attacking people and, while some can be deemed defensive attacks, many are not. Andy Peterson, a cougar-attack survivor, described his experiences of an attack that happened in 1998 in enough detail to allow us to clearly infer a predatory motive. Writing in the UK newspaper the *Guardian*, Andy tells how 'the mountain lion smashed right into my chest, knocking me off the trail. Its claws were in my knees, my neck, my chest. Its jaw stretched over my head from my hairline to the back of my skull'. The cougar was attacking the head and neck region and, from Andy's description of the attack, appears to have been trying to get its canines into a vertebral gap. The attack happened about 24km (15 miles) or so from the city of Denver in Colorado. Andy survived by gouging the cougar's eye deeply, pushing his 'thumb through the eyelid all the way to the muscle at the back of the eye'. He was carrying a small knife and was also able to stab at the cougar's head. Eye gouging and stabbing are not the usual prey responses and the cougar clearly had second thoughts. Andy was able to run away and escape, only to spend six hours in surgery to get patched up. His head wound alone required 70 staples to close it, and he had further wounds across his neck, chest, thighs, shoulders, arms and stomach.

Andy's story has an interesting ending. As he says, 'Almost a year later I received a phone call saying a female mountain lion had been cornered in a nearby garden. The lion had a one-inch scar on the top of her head and was missing an eye. I had the choice to have her shot there and then. Arguing that she'd just been defending herself, I persuaded the park rangers to set her free.' At the time of his attack, Andy had been hiking to clear his head during a time in his life when things were not going well for him. In his own words, 'I'd hurt people through being malicious and careless, and the lion hurt me through natural instinct. In a way, it had more of a right to live than I did. The whole incident changed my life for the better. I gave up drink and drugs, and found religion. I've been saved twice, once from the lion and once from my own self-destructive urges.'[4]

Andy's story is unusual, but it serves to highlight the often-complex relationships we can have with the natural world. Not everyone wants retaliation, and not every community wants predators removed. However, missing an eye is a fairly serious handicap for a predator that relies on vision and depth perception to find and hunt prey. Not every injured predator becomes a human-eater and not every predator that attacks humans is injured, but there is certainly enough of a relationship there to make you think twice about the wisdom of leniency. Were another attack victim to relate how they had fallen foul of a cougar with a missing eye and distinctive scar then things might look less rosy. Nonetheless, I respect Andy's decision and, despite having reservations, I would like to think I would have done the same in his position – but it would have been a tough call.

Cougar attacks are not especially common, but because they can happen in the United States and Canada they inevitably attract mass-media attention. Commenting on the closure of two schools in north California in 2021 provoked by the sighting of a cougar nearby, the Mountain Lion Foundation sums it up statistically by stating, 'You are more likely to drown in your bathtub, be killed by a pet dog or hit by lightning [than to be attacked by a cougar].'[5] As we've seen already with the perception of attacks for different species, while the overall risk may be very small indeed, the notion of being attacked and eaten by a wild animal is sufficient to make people lose touch with statistical reality. And while the Mountain Lion Foundation is correct in saying that the risk is extraordinarily small, it nonetheless remains the case that cougars can and do target people in predatory attacks. In 2008, Robert Nawojski was attacked, killed and partially eaten by a cougar in a wooded area near his home in Pinos Alto, New Mexico. Parts of the body had been buried for later consumption.[6] A search of US news for cougar attacks reveals a string of attacks and near-misses. 'Cougar attack: BC woman seriously injured, airlifted to hospital'; 'Mountain lion killed after mauling boy'; 'California: mother fights off mountain lion with bare hands to save five-year-old son'; 'Cougar stalks hiker'. These are just from 2021.[7]

An authoritative and comprehensive analysis of cougar attacks was undertaken by Paul Beier and published in 1991. During the period between 1890–1990 in Canada and the United States, he recorded 52 incidents and 10 fatalities. More than a third of the incidents Beier records happened in one area, Vancouver Island, off the coast of British Colombia. It isn't clear why this should be the case. Hans Kruuk, in his book *Hunter and Hunted*, suggests that there could be what he terms a 'cultural effect', a social learning process resulting in the local predator population coming to regard humans not as something to be feared, but as something to eat.[8]

Other commentators echo aspects of Kruuk's analysis. In 2004, US journalist David Baron wrote about cougars in North America, using the death of 18-year-old Scott Lancaster as a framing device. Lancaster was killed and eaten by a cougar that attacked him during the day while he was jogging on a hill near

Clear Creek High School in Idaho Springs, Colorado, in January 1991. He was the first person to be killed by a cougar in recorded Colorado history. In *The Beast in the Garden*, Baron relates how, before the attacks, scientists studying cougars in the area reported changes in cougar behaviour. The cats had lost their fear of people and were considering humans prey. Baron used the attacks as a vehicle to advance a thesis that in North America our relationship with the natural world, and with certain species, was changing. His journalistic work had produced many wildlife stories that were not about wildlife declines or conservation issues, but instead focused on certain species increasing in abundance and their encounters with people. He reported on alligators eating dogs, bears raiding rubbish bins, and deer entering gardens to eat flowers. In an interview with Colorado newspaper the *Vail Daily* in 2005, Baron recounts that he 'was struck that all over the country we're entering a new phase in our relationship with wildlife.[9] We're living in closer proximity to many of these animals than we have before. We're also bringing the animals back to abundance and they are moving in to where we are.' More predators, living in closer proximity to more people, might inevitably lead to more attacks, but are we really seeing an upwards trend in cougar attacks? Answering that question depends on what timescale we consider.

If we take a broad timescale then the answer seems to be yes, we are seeing a rise in attacks and in fatal attacks. In his analysis, Beier reports that there were more fatal attacks (six) between 1970 and 1990 than in the entire period of 1890 to 1970 (four). It is possible that this upward trend has continued. Many have noted that in recent years there seems to have been something of a spate of attacks, or at least, a spate of media stories reporting attacks. It is commonplace in articles reporting attacks to see words to the effect that 'we are seeing a rise in cougar attacks' and, while that might be true across the broad historical perspective Beier took in the 1990s, it is harder to be quite so confident about the trend over the past few decades. Commenting in US magazine *Outside* in 2019, Lynn Cullens of the Mountain Lion Foundation points out that when a dataset (fatal attacks by cougars) is so small ('fewer than two dozen' in the past 100 years) we don't really have too much to go on. A couple of attacks in one year, although shocking and tragic, is not enough for us to say that there is any upward trend.

Both Cullens and Baron converge on broadly similar explanations for why cougar attacks are happening and, interestingly, the explanation bears a remarkable similarly to some of those proffered for increasing tiger attacks in India and Nepal. The pattern is quite simple and I'll outline it for cougars, but you can easily replace species names and locations, and tweak dates, to make it more widely relevant (*e.g.* see wolves in Europe in Chapter Nine). When Europeans arrived in North America they saw cougars as a threat, and as competition for game species like whitetail and mule deer. Unregulated hunting, which included bounties for dead cats, led to dramatic declines in abundance and range. By the mid-1990s, cougars were isolated in a fragmented pocket of their former range. Conservation efforts then kicked in, and a combination of protection and active intervention have led to a recovery in numbers and an expansion of range.

The recovery of cougars in North America is yet another great example of how great we are at conservation when well-known, charismatic species are sliding rapidly in the wrong direction. But it is also another great example of how badly we can let things slide before we do anything about it. That uncomfortable fact notwithstanding, the increases in cougar numbers and range will inevitably lead to more cougars with no experience of humans potentially interacting with us in areas where previously big cats were absent. We should note, though, that it isn't as though people are wading through cougars to get to work. These are not animals that live at especially high densities and estimates of the number in North America are around 30,000. The picture is also more complicated than cougar range expansion because, as Baron noted in his book, humans have also expanded their range in North America. Since 1990, when Beier's analysis of attacks ends, 60 per cent of new homes in western states, where cougars are expanding, have been built in what is termed the 'wildland-urban interface'. Areas of natural beauty, with rugged wooded mountain scenery loaded with deer, are attractive to people but are also prime cougar country. Both species in this predator–potential prey relationship are expanding into each other's 'territory'. Humans have another aspect of their ecology not found in other species that influences the probability of coming into contact with cougars: recreation. The number of people visiting US national parks has been rising steadily, with the most popular park – the Great

Smoky Mountains National Park in Tennessee and North Carolina – increasing visitor numbers from just over 9 million in 2008 to more than 14 million in 2021. Some visitors may rarely leave their car or picnic site, but many do. The number of people hiking in national parks nearly doubled between 2006 (30 million) and 2020 (58 million).[10] Hiking, camping, climbing and other activities in the 'great American wilderness' mean that people are outside in potential cougar country in record numbers. This background of human expansion into wilderness makes the actual numbers of cougar attacks, fatal or otherwise, remarkably low, especially when we compare them with what is happening in Tanzania with lions or in India with tigers.

In addition to conservation interventions causing an increase in predator number and range expansion, there is another interesting parallel between cougars in North America and tigers in India. Cullens attributes cougar-related incidents in recent years to juvenile animals dispersing to new areas. These younger cats are not yet proficient hunters and, like the tigers Rajeev Mathew spoke to me about in Chapter Three, may be looking to people (and their pets) for easy meals. Older animals are far more likely to have developed an aversion to people. Cullens also outlines how the hunting of larger cougars in Oregon might be adding to the problem. By removing these more experienced cats, hunting is opening up opportunities for inexperienced juveniles to replace them. Mathew thinks that younger tigers grow out of hunting humans. Whether this is the case with cougars remains to be seen, but Cullens is definitive in her position, telling *Outside* magazine that 'The best way to prevent mountain lion attacks is to stop killing mountain lions.'[11]

As we saw with hyenas (see Chapter Six), cougar presence can have tangible human benefits. One potential benefit is cougar tourism, but the secretive nature of these cats makes encounters extremely unlikely. Nonetheless, the potential presence of a large predator, even unseen, enhances the sense of wilderness and may be a factor in drawing people to stay – and spend money – in some areas. It may also put people off, but the huge increase in outdoor recreation in recent years in areas where other large predators (notably black and brown bears, see Chapter Eight) are present suggests that there are always some people willing to take the chance. Another benefit is more indirect, but potentially more life-altering than an encounter with 'the ghost of the forest'. Wildlife ecologist Sophie Gilbert led a team that examined the benefits cougars may provide by killing deer and thus reducing vehicle accidents. Deer are a significant cause of road accidents, whether it is people swerving to avoid deer or hitting them (a deer–vehicle collision or DVC) and subsequently losing control. The team's modelling approaches 'revealed that cougars could reduce deer densities and DVCs by 22 per cent in the eastern United States, preventing 21,400 human injuries, 155 fatalities and US$2.13 billion in avoided costs within 30 years of establishment. Recently established cougars in South Dakota prevent US$1.1 million in collision costs annually.' These benefits are substantial, and go far beyond the raising of a few bucks through potential tourism opportunities. There are hundreds of people who owe their lives to cougars and thousands more who have cougars to thank for the fact that they can still walk unaided.[12]

There is no doubt then that cougars can and do attack people. Many such attacks are clearly predatory and there are well-documented cases of victims being eaten. It is foolish to suggest that these animals pose no risk at all, but equally the number of people being attacked is extremely low and your chances of even seeing a cougar, let alone ending up dead due to one, are vanishingly small. The low number of attacks, both in absolute terms and as a proportion of the people who will be in cougar country by virtue of where they live or spend their leisure time, illustrates that we are not really a major blip on the prey radar. Remember, around Bardiya National Park in Nepal in the last half of 2020 and the first quarter of 2021, 10 people were killed by tigers. That is the same number of people who have been killed by cougars across the whole of North America in the 100 years between 1890 and 1990. In the UK press, the spate of tiger attacks in Nepal barely got a mention, whereas cougar attacks are frequently covered. Even cougars 'stalking' people get huge amounts of press attention in a country thousands of miles away.[13] Like Gotz Neef surviving a lion attack in Botswana (see Chapter Two), the only real headline here is that a Westerner getting stalked by a cougar for – as the headline puts it – 'SIX MINUTES', or being mauled but surviving a big-cat attack is far more newsworthy than, say, a large number of Indian or Nepalese people being stalked, killed and eaten by tigers – and that isn't right.

The jaguar

The cougar shares a large part of its southern range with the continent's other big cat, the jaguar. It is the only extant *Panthera* species that lives outside of Africa and Asia, and is the third largest member of that genus, after the tiger and the lion. Superficially similar to the leopard – a big cat we'll meet in some detail shortly – the jaguar is a bulkier and more solidly built animal than its African/Asian counterpart. Both cats have a coat marked with dark spots on the legs that grade into rosettes on the body, but the jaguar's rosettes are more widely spaced and have a distinct spot in the middle. Very big males can approach or exceed 100kg (220lb), especially in captivity, but typically these cats weigh about 60–80kg (132–176lb), with males being 10kg (22lb) or so heavier than females.[14]

Historically, jaguar range was far more extensive than it is now, which is a phrase I am getting tired of typing. The jaguar once ranged over much of South America (with the exception of the very south and the high Andes), up through all of Central America and Mexico and into the southern United States. Its range across this region has reduced and become considerably fragmented over the past century or so, losing ground in the southern United States and northern Mexico, as well as large areas of Brazil and Argentina. Habitat loss and fragmentation to forestry and agriculture, together with (now illegal) hunting for skins and retaliatory persecution for killing livestock have taken their usual toll. Road building is also a concern because it facilitates the third of the familiar triplet of threats – prey-base depletion. A study in Ecuador showed that road building had cascading effects on wildlife. Roads allow hunters easier access to previously hard-to-access areas. An increase in hunters leads to a decrease in prey and, in this study at least, an up to 18 times lower density of jaguars.[15]

The jaguar's lifestyle has strong similarities with the tiger's way of doing things. These are solitary animals, with groups only occurring when females have cubs. Females generally have smaller home ranges than males, with male home ranges overlapping with several females. Jaguars are forest cats, preferring habitat with dense cover across much of their range, although they can also be found in wetlands and grasslands. They tend to be associated with water and are keen and strong swimmers. Jaguars are active mostly at night, and are generally considered nocturnal and crepuscular (active at dawn and dusk), but in some areas, especially densely forested areas, they hunt during the day. This is the case in the Amazon rainforest and this activity pattern is something that will feature when we consider negative interactions with people.

Being animals of dense cover and active often at night makes jaguars hard to see and difficult to study. My first sighting of one well illustrates this point. I was studying stingless bees in Yucatán, Mexico (the bit that sticks up into the Gulf of Mexico) in the early 2000s and, during a break in some experiments, I took a week to drive south and explore the Calakmul Biosphere Reserve in Campeche. Driving into the reserve down an access road though heavy forest, I turned a corner and saw, around 30m (98ft) ahead, a young male jaguar strolling down the road towards me. I stopped the vehicle and waited, expecting the cat to dart off into the forest, giving me a thrilling but brief sighting of my first 'in the wild' big cat. In fact, the jaguar kept coming and stopped a few metres in front of car. It stayed there for 10 seconds or so, then continued past the car and around the corner I had just driven.

An unforgettable sighting, and I was buzzing as I got to the entrance building around a kilometre further on. Eventually realising that the local word for jaguar was '*tigre*' (tiger), I excitedly told the staff in my bad but just comprehensible Spanish that I had seen '*el tigre*' down the road. I duly filled in a sightings book that had no mention of jaguars, 'tigers' or any big cats that I could find. I later learnt that an American biologist was studying jaguars at the site and, despite many hours in the field tracking and collecting scat (some most likely from the male I had seen), had never seen one. I got very lucky and I doubt I will see another wild jaguar. They are elusive, secretive cats.

Jaguars are powerful predators that rely on meat for nutrition. Their prey spectrum is broad, ranging from 1kg (2.2lb) rabbits to 130kg (287lb) tapirs. Those are extremes and they mostly favour prey in the mid-range, including capybara (the world's largest rodent), caiman (a nice example of predator–predator predation), white-tailed deer, wild pigs and tamandua (a small species of anteater).[16] They have a powerful bite, which allows them to tackle heavily armoured prey like turtles and tortoises, and to deal swiftly with prey that could bite back, especially caimans. Unlike other big cats, when tackling mammal prey the jaguar doesn't bother with any subtlety involving vertebrae, although it does adopt that approach with caimans. Its approach with mammals is more direct, driving its canines straight through the skull of its prey and into the brain. Jaguars aren't fans of pursuit, preferring instead a stalk-and-ambush approach, much like the tiger. And, like tigers, larger prey is often dragged to dense cover to be consumed, which tends to happen in a pretty regimented fashion. Jaguars first remove the digestive tract, then begin eating, starting on the neck, chest, heart and lungs, before moving on to the shoulders.[17]

Jaguars are large 'big cats', powerful and predatory, and have a hunting ecology strongly reminiscent of tigers. They are capable of taking prey weighing up to 130kg (287lb), which comfortably includes most people. So, jaguars certainly seem like excellent candidates for being human-eaters. Balanced against that, these tend to be secretive cats that prefer dense forest, but the same could be said of tigers and as we've seen this doesn't stop them from taking people. The key to the tiger attacks in places like Bardiya National Park and the Sundarbans (see Chapter Three) was not so much tiger behaviour but human behaviour. In both of those locations the drive for people to enter forests to collect resources put people into potential contact with predators. I grew up believing that jaguars never attacked people, and vividly remember reading in the sort of nature book to which I was addicted as a child, that the reasons for this were that jaguars and humans hardly ever come into contact. This may have been a comforting narrative at the time but, sadly, it isn't entirely true.

Jaguars are known to attack, kill and eat humans. Their technique, jamming large canines through the skull into the brain, leaves little room for survival and attacks are often fatal. One non-fatal attack, reported in 2015 in the *Western Journal of Emergency Medicine*, describes an incident involving a three-year-old girl called Jasmine Joseph from the Isseneru Village, in the remote Cuyuni-Mazaruni region of Guyana, near the Mazaruni River in dense forest. It was 27 December 2013 and the girl was with her mother Desiree Joseph who was bathing and washing clothes, a human behaviour pattern we have come across before with victims of crocodilian attacks in Chapter Four. As the mother was washing clothes, a large adult jaguar hunted the daughter in a stereotypical sequence of events that characterises their attacks. The cat, which had stalked close to the girl unseen, pounced on her from bushes, clamped its jaws around her head then dragged her about 20m (66ft) into bushes.[18] The girl's weight is given in the report as 24kg (53lb), which I think is a mistake. The average weight of a three-year-old girl, according to charts published by the Royal College of Paediatrics and Child Health in the UK and based on World Health Organization data, is about 15kg (33lb).[19] A 24kg (53lb) three-year-old would be in the 99.6th percentile, even if they were almost four. The report is in an American journal and elsewhere uses imperial measurements. I suspect the child weighed about 11kg (24lb), which to a jaguar is small prey. The girl was the equivalent of a large tamandua or a very small capybara (which have an average adult weight just short of 50kg/110lb).

The screams of the child, who was remarkably still alive despite being dragged 20m (66ft) by the head, brought other residents to the location. That they were alerted by

screaming indicates that the attack happened close to the
village and therefore close to regular human activity. This
was not an attack a long way from home, although the
village itself is in a very remote region (a description that
could be applied to much of Guyana). The residents are
said, in a local press report, to have intervened 'with a
cutlass'.[20] Cutlass is the word used throughout that region
for a machete. The jaguar, reported as a suspected puma
by the local press, was killed apparently by being shot
(although this detail in the scientific write-up is supported
by a reference to a local press report that doesn't mention
a shooting). The girl, suffering extensive head injuries,
was air-lifted to hospital and put on life support.

As might be expected given their secretive ecology,
jaguars are not commonly seen by residents in this region.
Despite this, Jasmine's grandmother Agatha had
witnessed another attack by a jaguar in the same area just
a month before the attack on Jasmine. As she described it
to the Kaieteur News, 'Last month (late November last)
the animal jumped on her [the victim] and scraped her
on her foot, but I hit it and it get away.' It seems very
possible that Jasmine was attacked by the same animal.
She was less fortunate and suffered extensive injuries.
Multiple deep lacerations were present over her head and
face, and the paper contains graphic images of the
injuries. One of the lacerations was associated with a
fracture of the skull so extensive that it was clearly visible,
while a CT scan showed more extensive fractures along
the left frontal bone involving the eye orbit. In the
operating theatre yet more skull injuries were revealed
along with other facial and soft-tissue damage. Jasmine
was discharged 22 days after admission, with healing

wounds, some scarring (although the surgeons had done an outstanding job) and some ptosis, a drooping eyelid likely caused by local nerve damage.

Jasmine really was very lucky. She would certainly have died were it not for rapid intervention from nearby residents, an airlift to a major hospital, and skilled surgical intervention. The same applies to a 17-year-old man attacked by a jaguar in the north Pantanal (an extensive wetland area) in Mato Grosso state, Brazil, in 2010. A jaguar jumped out of a ravine and on to the boat in which the victim was returning from a fishing trip. Biting the victim on the shoulder, the jaguar tipped him into the water. The cat surfaced shortly afterwards with the victim's head in its jaws, the now familiar jaguar 'death bite' fully in play. The boat's skipper brought an iron pipe to the situation and the jaguar released the victim before escaping. The head wounds are dramatic, with brain and bone tissue losses clearly evident from the photographs published in the medical report detailing the attack.[21] The victim suffered other facial, arm and torso injuries. Thanks to some rapid first aid, and the good fortune of encountering another boat with a doctor on board, the victim survived but was reported to have memory deficits. Another fisherman in Mato Grosso was less lucky. In June 2008, the 21-year-old man was sleeping in a tent at night when a jaguar entered, killed him and dragged the body 60m (197ft) into the forest. The jaguar was disturbed before the body was consumed, but the style of the attack and the dragging of the body clearly indicated a predatory motive. This attack was confirmed later as the first official record of a human killed by a jaguar in Brazil.[22]

Jaguar attacks are extremely rare and fatal attacks are even rarer. My childhood nature book was more or less correct. But, as with other human–predator interactions, it is entirely possible that attacks could become more common. Jaguar habitat is decreasing and fragmenting, which inevitably leads to an increased likelihood of humans and jaguars coming into contact. Prey depletion is also a strong factor. Species that jaguars like to eat are also eaten by people, and local hunting pressure on capybara, tapirs, peccaries and caimans is likely to lead to jaguars seeking other prey, including domestic animals, livestock and possibly people. Some authors have suggested that ecotourism may also play a negative role in human–jaguar interactions. Excursions and viewing tours in the Pantanal have involved baiting jaguars using food and attracting them further by mimicking the calls of females in the breeding season. These types of interactions have at least two effects. First, jaguars become habituated to some extent to the presence of humans. Second, jaguars come to associate people with feeding. Ultimately, neither may be good associations for either species. Jaguars may not have been a threat in the past, and are only an unlikely threat now, but we may still need to change our behaviour as we progressively change their environment if an escalation of conflict is to be avoided.

Leopards

Leopards are perhaps my favourite of all the big cats. Lighter in build than a jaguar, the dark rosettes covering a leopard's pale-yellow coat tend to be smaller than those covering its Latin American cousin and lack any central spot. As in the jaguar, the rosettes grade into solid black spots on the head and legs. This spot and rosette pattern, like the saying tells us, doesn't change. It also varies, sometimes greatly, between individuals. This lets us distinguish and identify individual leopards, so long as we get a clear view.

Getting that clear view of a leopard can be tricky. These cats are masters of moving through an environment without being seen. I once heard a story of a man who lived for 20 years beneath a large rocky outcrop known in South Africa as a *kopje* (pronounced koppie). Most days the man would see or hear some sign of leopards living in the area, prowling the *kopje* and coming close to his house. He would hear a male leopard making its distinctive rasping, roaring sound, known as sawing due to its resemblance to sawing wood. He would see leopard tracks on his daily walks around the property and discover kills in trees. Leopards, like some other big cats, drag their kills to a safe place to eat and store but, unlike other cats, leopards are able to drag prey considerably heavier than themselves up a tree, stashing it in the crook that forms where a branch meets the main trunk. And yet for 20 years the man never once saw a leopard.

I have been more fortunate that the man who lived below the *kopje*, although I also saw and heard numerous signs of leopard before I finally saw one. That sighting, of which I can remember every second, occurred at the end of an already magical day. I spent the morning being up close, touching in fact, a white rhino that I had watched being darted from a helicopter to be treated for a suspected gunshot wound. I thought the day wasn't going to get much better when, as the saying goes, the bush was kind. My first leopard was a big male, who I picked up from his movement as he walked about 40m (131ft) away through scrubby grassland.

My first sighting underlined two things for me, First, that leopards are large, powerful animals. Because they are quite a bit smaller than lions or tigers it is easy to think of them as 'small' big cats, but a very big African male can weigh more than 90kg (198lb). Their build is less heavy-set than a jaguar's, but these are muscular animals, as their tree-dragging behaviour shows. The second thing I realised very quickly was just how much a part of their landscape leopards are. They blend in and disappear. Despite the cat being relatively close, in fairly light cover – and moving in a predictable, direct line – I frequently lost sight of him. He just seemed to vanish completely and then fully reappear a moment later. The leopard's reputation for stealth is well deserved.

Leopards have a broad geographical range, indeed the broadest of any cat species. It will come as no surprise to read that this range is smaller than it once was (perhaps just 25 per cent of historical range) and has become fragmented in recent times through human activity. Nonetheless, leopards can be found across a broad part of the globe, including much of sub-Saharan Africa, through western and central Asia, the Indian subcontinent, Nepal, Bhutan, Myanmar, and down through Malaysia and Indonesia. They are also found in parts of China, Russia and Turkey.

As we might expect with such a broad geographical range, a number of subspecies are identified. As we might also expect, after our brush with subspecies in

Chapter Three, the situation can be a little confused, and confusing. Leopard taxonomists went on something of a subspecies rampage, recognising 27 based on physical variation and geographical location. Several of these subspecies were based on just a few skins or skull specimens, which is problematic when you are assessing subspecies based on population distinctiveness from other populations. Animals can vary considerably within a population, so to account for that within-population variation while determining between-population variation requires more than just a few incomplete specimens. In 1996, the leopard subspecies situation was addressed using molecular genetic approaches, examining genetic differentiation and divergence between samples from 60 leopards across their range.[23] This analysis resolved the 27 subspecies into eight, extended to nine by another analysis in 2001.[24] Further analyses followed using improved genetic techniques and better sampling. The overview remained broadly similar, but some disputes still remain. In 2017, a lengthy report entitled 'A Revised Taxonomy of the Felidae' produced by the Cat Specialist Group of the IUCN, opted for what it terms a conservative approach, identifying eight subspecies across the range from Africa to Indonesia.[25]

As predators, leopards are usually considered opportunistic and, as befits their broad geographical range, their prey spectrum is wide. They have been known to take prey as large as eland (the largest antelope species, topping more than 900kg/1,984lb) and young giraffe, and as small as rats. They will hunt primates and are partial to other carnivores including jackals, but their favoured prey tends to be small antelope (such as impala and duiker in Africa) and deer (like the chital in India). Leopards also enjoy eating dogs. This habit is well known in Africa and is on the rise in India where, as we will see, feral dogs may be supporting increasing populations of urban leopards. Leopards will take cattle and donkeys, and do so across their range, but they are far more likely to predate smaller livestock like goats and sheep, and prefer calves over adult cattle.[26]

Livestock losses to leopards can be considerable or, more accurately, can impose considerable costs on those whose livestock are taken. In a study of livestock predation in Bhutan, 274 households reported the loss of 76 domestic animals to predators, of which leopards accounted for 40 (53 per cent).[27] This amounts to a 2.3 per cent loss, which may seem relatively minor when viewed from a distance. However, when the microeconomics of villager life are accounted for, the losses are considerable. The average annual household cash income in the year 2000 was estimated to be US$250. Taken across all households (including those that didn't report losses) each lost an estimated US$44.72 or around 17 per cent of the annual household cash income. This amount, almost a fifth, is a hard loss to bear, but if only the households that reported livestock losses are considered the costs become immense.

If we consider the mean losses for only those households that reported losses, the figure was US$211, or around 84 per cent of the annual household cash income. More than half of these losses were caused by leopards. In part, losses may be exacerbated by the tendency of leopards to indulge in what is known as 'surplus killing' or 'henhouse syndrome'. Surplus killing describes situations when predators (like the archetypal fox in a henhouse) faced with a high density of prey kill a number far in excess of what they can realistically eat. This is not an uncommon behaviour. A study of surplus killing in the Cape province of South Africa published in 1986 reported 104 incidents of livestock being killed by leopards, of which 76 (73 per cent) involved the killing of two or more animals. Some 64 (62 per cent) of the killing incidents resulted in between two and 10 prey animals being killed and one incident resulted in the killing of 51 sheep and lambs. This study suggests that, at least in some areas, surplus killing may be the norm when the opportunity presents itself.[28] I have heard many such stories from people living in South Africa and Namibia. It isn't just African leopard behaviour; in 2020, a leopard in India got into a shed containing livestock and killed at least 47 goats, injuring a further 18, with 25 unaccounted for.[29] Such incidents, with their losses spread out over a community, might appear as trivial relative losses, but when considered at the level of the household directly affected they are likely to represent a devastating near-total annihilation of wealth. It is easy to see this by comparison with a more familiar scenario. The median household income in the UK in the year between April 2019–March 2020 was, according to the Office for National Statistics, £29,900.[30] Now imagine that the presence of a predator in the area took away £25,116, leaving you with just £4,784 (the 84% household loss described above). I don't know about you, but if I were faced with that situation I am not so sure I'd be looking at leopards with the same doe-eyed affection and admiration as I do from the comfort of an air-conditioned vehicle.

Livestock losses impose major costs, which themselves can be indirectly life threatening or life limiting. But given their propensity to hunt animals that are more or less the same size as a person, we shouldn't be surprised that leopards can also impose more direct costs. For one of the best-documented cases we can once again return to the writings of Jim Corbett, a man we last saw guiding Princess Elizabeth around the Treetops Hotel in Kenya at the point that she became Queen Elizabeth II. The so-called Leopard of Rudraprayag was a male that killed more than 125 people between 1918 and 1926. The leopard made its first kill in the village of Benji in the Rudraprayag district in Uttarakhand State in northern India, and in the eight years that followed its depredations caused considerable fear, for good reason. A popular road between two Hindu shrines passed through the area and travel at night became a highly risky undertaking, although people didn't need to go out at night to become prey. The leopard would attack residents in their homes, taking victims from their beds. A number of attempts were made to kill the leopard, including some by well-known hunters, but it fell to Corbett to sort out the problem posed by what he described as 'the best-hated and most-feared animal in all India'. It took him 10 weeks, but eventually he shot the leopard in May 1926, ending its reign of fear.[31]

When it came to big cats taking people, Corbett was a strong proponent of the 'old and/or injured' hypothesis. It certainly seemed to apply to many of the tigers he shot. However, the Leopard of Rudraprayag was different. By the time it was shot it was an old male (and a large one), but it had begun its reign as a human-eater eight years previously when it was young. The leopard showed some signs of injury, sustained from earlier botched hunting attempts, but these were inflicted later in its life and had healed. Corbett came up with an alternative hypothesis for the leopard's dietary preference – and one that harks back to earlier discussions about hyenas. It was Corbett's contention that the leopard had gained a taste for human flesh by feeding on bodies left unburied during disease epidemics. As he writes, in the *Man-eaters of Kumaon*, 'A leopard, in an area in which his natural food is scarce, finding these bodies very soon acquires a taste for human flesh, and when the disease dies down and normal conditions are established, he very naturally, on finding his food supply cut off, takes to killing human beings.' The first victim was taken in 1918, which coincides with the influenza pandemic known as Spanish flu. The disease swept through India that year, killing an estimated 18 million people and most certainly leaving many unburied.[32] The idea of a 'modified environment' influencing human predation is one to keep in mind; it will come up again in Chapter Nine when considering the risk of wolf attack.

We can never be sure of the history of the Rudraprayag leopard and correlation does not imply causation. There is, however, another interesting parallel between disease outbreaks and leopard–human predation in northern India, again involving Corbett, which lends weight to the idea. The first human-eating leopard Corbett shot was the Panar Leopard. This cat was by far the more prodigious human predator, killing at least 400 people before Corbett killed it in 1910. The start of those attacks coincided with what Corbett described as 'a very severe outbreak of cholera'. The pattern fits, but as we will see shortly it doesn't require a disease outbreak to stimulate a prey shift in leopards.

Leopard attacks were rare at the time of the Rudraprayag and Panar incidents. Tigers were by far the most feared predators, certainly accounting for most of the human-eating cats Corbett shot. That is not the case these days. According to a report in the *Indian Express* in 2018, of the 182 big cats (tigers and leopards) declared to be human-eaters in the state of Uttarakhand in the 13 years between 2005 and 2018, 166 (91 per cent) were leopards.[33] In the first six months of 2018, 11 big cats were declared a threat to life and seven of these were leopards. There were an estimated 60 leopard attacks a year in 1998–2012 in Uttarakhand. The population of Uttarakhand in 2001 was just under 8.5 million, rising to just over 10 million by 2011.[34] The current population of London is just a fraction under 9 million, so Uttarakhand is more or less comparable to London in terms of size of population. If a fox attacks someone in the UK (an incredibly rare event that has never resulted in a death) it makes headline news, complete with lurid language and pixelated images of what generally turn out to be minor injuries.[35] Just remember the next time you read about such a 'horrific attack', that in just one state in India there will probably have been more than one leopard attack that week.

As you can imagine, the injuries from non-fatal leopard attacks are very far from minor. Leopards are particularly dangerous animals when cornered. They will charge quite fearlessly at whatever stands between them and escape. Launching themselves at their tormentor they will often engage the shoulders of a person with their front legs, driving claws deeply into flesh, while raking up and down with their powerful back legs. Such an attack pretty much guarantees serious, life-threatening injuries, and disembowelment is a genuine possibility. Quick and brutal, a defensive leopard strike can be just as fatal as a deliberate predatory attack, which tends to target the throat and neck region. Even if the initial defensive strike doesn't cause the victim to bleed out, without medical attention and antibiotics it is very likely that fatal infection will set in. The take-home lesson here is, never corner a leopard.

Official figures for fatal leopard attacks are almost certainly underestimates, for all the reasons we have explored in previous chapters. With that in mind, in the years between 2000 and 2007, just in the states of Uttarakhand and Gujarat, there were 289 officially reported human fatalities from leopards.[36] During this period both humans and leopards were increasing in numbers, and human–leopard conflict was becoming a serious problem. One major factor identified in a workshop held in New Delhi in 2007, aiming to develop a 'pragmatic management policy for human–leopard conflicts in problem areas' is the adaptability of leopards and their ability to live and thrive in human-dominated landscapes. Their diet is diverse and their predation more opportunistic than that of many other predators (with the exception perhaps of spotted hyenas, see Chapter Six). As their extensive geographical range and habitat diversity show, leopards are also able to tolerate many different ecological conditions. Overall, leopards are, at least to some degree, preadapted to live alongside us. In a populous country like India, this greatly increases the possibility of human–leopard interaction and subsequent conflict.

The rise of leopards in human-dominated landscapes raises issues for conservation that are also valid for other species. Protected areas, like national parks, are an important focus for conservation efforts, but in most nations they are a small proportion of land area. In India, they amount to around 5 per cent. This inevitably means that much of the land available for species to inhabit is unprotected and in many cases may well be human dominated. Understanding how species live in these areas (assuming they do) is clearly important if we are to understand how to conserve them. One attempt to do that was a study in western India that used camera traps to determine the presence of carnivores (primarily leopard and striped hyena) in agricultural landscapes with relatively high human populations.[37] The results were surprising. In an agricultural landscape, devoid of wildlife prey and very far from being wilderness, they found large breeding populations of both carnivores, with leopard approaching five adults per 100km^2 (39 square miles), a relatively high density. Some predator species can do rather well in highly modified landscapes.

Of all landscapes, the urban environment is the most modified of all, but even here leopards have found a niche. Unlike lions and tigers, leopards regularly take small prey, so have a prey spectrum perhaps more closely aligned with jaguars. Like jaguars, leopards are secretive, solitary animals, but unlike jaguars they are more at home around human habitation. It is here that their broad dietary spectrum pays dividends. The past few decades have seen the rise of the 'urban leopard', a phenomenon seemingly driven in part by the presence of large numbers of feral and free-ranging dogs. India has a major dog problem. There are an estimated 60 million dogs that are not under meaningful human control and these dogs have to eat. Dogs are pack animals and, when acting together, they can easily take down prey the size of deer and antelope (see Chapter Nine). This causes serious conservation concerns in many areas, and the control of these dogs is another topic where animal-rights agendas can clash with conservation goals. For the purposes of this chapter, though, these dogs are simply leopard food.

The reliance of leopards on domestic animals was demonstrated by a study of leopard scat undertaken in the state of Jaipur in north-west India in November 2017–April 2018. The authors, led by Swapnil Kumbhojkar, analysed 132 leopard scats and found that domestic animals occurred in nearly 90 per cent of them. By far the most common species to appear was the domestic dog (44 per cent), with domestic cats (13 per cent), goats (16 per cent) and cattle (15 per cent) trailing some way behind. The authors concluded that the presence of domestic prey is sustaining a local population of 25 leopards, with feral dogs being particularly vital. Feral dogs attack people and spread rabies, and are identified as a public health problem in India. Like the hyenas of Ethiopia reducing potential disease burden by scavenging infected carrion, by taking these dogs off the streets the leopards of India 'could be considered as suppliers of a service to the human population amongst whom they thrive'.[38] A study of leopards in Mumbai quantified the benefit of urban leopards to people, concluding that their predation of dogs could be saving up to 90 human lives per year, depending on various assumptions, with knock-on conservation benefits through the reduction of predation of wildlife by dogs.

The Jaipur study also identifies a cost to leopards of predation on dogs, which is that it 'potentially exposes the leopards to the canine distemper virus' (harking back to the lions of Gir National Park in Chapter Two). That may well be true, but a bigger issue is that dog predation likely exposes leopards to people. With their urbanisation, facilitated by readily available food, leopards become habituated to people, less wary around them and come to associate areas where people live with prey. Coupled with increases in leopard population, this sets the scene for urban leopard conflict to take hold. Rajeev Mathew, who you will recall was collating data on tiger attacks in India (see Chapter Three) is also collecting data on leopard attacks. He describes the role of feral dogs in fuelling the increase in leopard populations and subsequent human–leopard conflict in semi-urban, urban and even city outskirts as 'massive!'.

I asked Mathew about the situation so far with leopard fatalities in 2021. He told me that 'as of July a total of 63 attacks are recorded'. Remember, Mathew is collating his data from newspaper reports and other records trawled from all over India; there is no central database recording these attacks. Mathew's approach is thorough, but it has a fatal weakness that means Mathew's figures are, as he puts it, likely 'far fewer than the actual number'. The problem Mathew faces is 'that leopard and tiger attacks rarely make national news any more … most go unreported. I can sometimes find a mention in some obscure local vernacular newspaper or some local news but with so many languages, it can be impossible to decipher news, printed or otherwise.' When people ask, quite understandably, how many people a year get killed by a specific predator, the really honest answer is that, for most species, we simply don't know.

Leopard dietary preferences give us some clues as to who is most likely to fall victim. They prefer smaller prey, with feral dogs being a favourite, so smaller humans are probably most at risk. Mathew puts it more plainly, suggesting that, 'Anyone under four-and-a-half feet tall is at a risk of being taken by a leopard.' This makes children particularly vulnerable and when victims are children it can sometimes result in attacks being reported more widely. In 2018, in what became a widely reported incident, a three-year-old boy, Elisha Nabugyere, was killed and eaten by a leopard in Uganda. A two-year-old was killed and eaten in Kruger National Park in 2019; two children were attacked, and one eaten in India; a six-year-old French boy was killed in Tanzania.[39] Most attacks, though, remain unreported even, as Mathew observed, in very local media.

Conservation front lines
The 'other' big cats bring up a fascinating range of issues pertinent to our relationships with predators.

Some of these issues are predictable, but some are rather surprising. Clear and surprising benefits of predator presence are shown both by cougars saving us from road traffic accidents and by leopards saving people from rabies by eating dogs. But these benefits are balanced, and may be overbalanced, by the costs imposed through livestock depredation and human attacks. The superficial similarities of jaguars and leopards are rapidly overshadowed by their differences, with leopards moving into – and seemingly thriving within – human-dominated landscapes, but jaguars suffering as landscapes become converted and habitats fragmented. In both cases, though, our effects on the landscape have resulted in increased contact and conflict with these species. Their ability to live alongside people means that leopards are increasing in numbers in India but causing large problems for people that we should not ignore. And yet we largely do. I have read very little in the Western media about the rise and associated risk of urban leopards in India, and yet articles saying that leopards are on the brink of extinction are two a penny. Formal protection has led to an encouraging rebound in cougar numbers in North America, which has caused very few attacks, but such incidents tend to dominate media coverage internationally because of the country in which they occur. If toleration and conservation of predators comes from understanding and balancing the cost–benefit ratio in more favourable ways, then we also need to balance our appreciation of the problems faced by people across the world as they try to live alongside animals that can, and do, kill and eat them.

Ultimately, the difference in attention paid to the odd, usually non-fatal, cougar attack in the United States and the near-daily deaths that occur in India is reflective of a far larger conservation issue. As we saw with tiger conservation in Nepal (see Chapter Three), and as can be seen in many successful conservation programmes globally, conservation works when it *involves* communities and fails when it is *imposed* on communities. Increasingly, it is becoming clear that if we want to conserve biodiversity then we must start listening to – and appreciating the perspectives of – those who live alongside it. If we don't we will achieve little or nothing. This is especially the case when it comes to conserving animals that cause conflict. But I think a bigger and more immediate issue is not just that we aren't listening to these communities, but that many people don't even realise such communities exist. Conservation is seen as something that is done to 'wilderness' for the benefit of 'wildlife'. Given the media depiction of wildlife and conservation, with a focus on glorious, pristine landscapes showcased in glossy documentaries – with fences, roads, powerlines and people skilfully avoided – this is hardly surprising. But with hyenas on rubbish dumps, lions in villages and leopards in towns, the number of communities on the front line of conservation is only getting bigger and their voices cannot be ignored forever.

CHAPTER EIGHT

Bears

I just did a search online for 'wilderness'. The images in first, second and fourth place were North American mountain scenes. The third image was of a wolf and the fifth was of a bear. For many, at least in the northern hemisphere, notions of 'the wild' seem inextricably linked to the predator species that are the focus of this and the next chapter. Bears (in all their diversity) have something of a bad reputation when it comes to interactions with people, despite the appeal of teddy bears, the musings of Winnie-the-Pooh and the good-natured mischief of Yogi. The fact that the North American expression for being fully prepared for any eventuality, especially a confrontation or challenge, is to be 'loaded for bear' tells us much about how these animals are regarded. But do bears really belong in a book about species that eat us? The answer is yes, but in a more qualified way than most of the species covered so far.

The bear necessities
Although not an especially large group, with just eight species, bears are more diverse in different ways than you might think. As a group of species they are also very widely distributed across the world, which of course gives many of them a high potential for coming into contact with people. With five large claws on each paw,

some formidable teeth, and an overall size and build that perfectly suits the word 'intimidating', those interactions may not always go well for person or bear. That said, several species of bear can be dealt with quite swiftly because they pose little or no threat to us in terms of predatory attacks. However, that is not to say they are completely harmless.

The obvious species to dispense with first is the giant panda. Pandas have long caused taxonomists a headache. This is well illustrated by the fact that, at one time, pandas were put in the same group as raccoons. This is hard to get your head around, I'll admit, but it makes some sense when we look at the evolutionary history of bears and raccoons, especially with the benefit of genetic analyses. These two groups are actually reasonably closely related and pandas share enough features with raccoons to have steered some early taxonomists in that direction. Pandas have also been aligned with the red or lesser panda, which looks like a cute cross between a ginger cat and a raccoon. In fact, the French word *panda* (the origins of which are unclear) was first applied to the red panda, with the far larger black-and-white species later getting the prefix 'giant' to distinguish it from its smaller assumed cousin. Again, an association that is initially hard to fathom, this presumed kinship was driven by the fact that red pandas and giant pandas share a very unusual anatomical feature. They both have an opposable 'pseudo thumb' that forms from the radial sesamoid bone of the wrist and which allows them to hold vegetation while feeding.[1] If you've seen a giant panda clutching a pawful of bamboo then you get the idea. Such a similarity, an unusual and seemingly identical anatomical feature, strongly suggests a shared common ancestry and for this reason these species were lumped together. In fact, and remarkably, the red and giant pandas' 'thumbs' are an example of convergent evolution. Convergent evolution is when distantly related species independently evolve similar features to overcome similar problems. It can be seen at work in the spines of hedgehogs and porcupines, and in the shapes of dolphins and sharks. After all the shunting around, though, the giant panda is now firmly considered a bear.

Giant pandas are big animals. A male can weigh more than 150kg (331lb) and although females are much smaller they can be more than 100kg (220lb). They have that burly 'bear build', reflecting considerable physical strength. Their skull comes furnished with the teeth of a carnivore, the large canines and sharp incisors being driven together with a bite force comparable to that of other bears.[2] Pandas, though, are famously vegetarian, eating bamboo, so as a food item we hold no appeal to them. Do not be fooled into thinking that pandas are just big, sad-looking cuddly toys, though. A paper published in 2014 with the title 'Three cases of giant panda attack on human at Beijing Zoo' does exactly what it says on the tin, and the photographs of the victims' injuries show the deep, life-changing wounds with which we have now become horribly familiar from other large carnivore attacks. The authors conclude that giant panda attacks are rare (and for once I think we can take that word at face value) and likely occur when pandas are 'infuriated or frightened'.[3]

The second bear we can eliminate from our line-up of potential human-eaters is Paddington,[4] by which I mean the spectacled bear, or Andean bear, of South America (including parts of darkest Peru). A mid-sized bear, it is around the same size as a giant panda and, like a giant panda, it is vegetarian. Or at least mostly vegetarian. Its diet is actually incredibly varied and includes delicacies like palm hearts, corn and berries, as well as less delicate choices such as tree wood, rodents and insects.[5] These bears aren't looking at us as food and neither are they especially aggressive towards us.

A study in 2005 of spectacled bear–human interactions surveyed a wide range of different people living in mountain regions of Colombia where bears were present, with the aim of collating and understanding interactions and potential conflicts.[6] Bears were reported crop-raiding and preying on cattle, but there were no reports of any injuries sustained by people as a consequence of bears, or of any unprovoked, predatory attacks. There were, however, two accounts of bear aggression. In one case, young children were watching a bear 'at close range without incident', but were then chased when the bear became aggressive. In the other case, a pig farmer had what is described as 'an aggressive encounter with a bear while defending his pigs', but neither farmer nor pigs were harmed in this incident. Given that these bears can take larger prey, and sometimes do, it is not inconceivable that a child or perhaps weakened adult *could* fall prey to one, but there is no evidence that this has happened. Paddington is, like so many carnivores, in far greater danger from

deforestation, persecution and habitat conversion to
agriculture (including in this case clearing land to
cultivate coca, the crop that is turned into cocaine) than
we are from him.

The final bear species we can eliminate as a human-
eater is the sun bear. We've met one before, inside the
belly of a large python in Chapter Five. The smallest of
bears, rarely exceeding 60kg (132lb), the sun bear lives in
the tropical forests of South East Asia where it feeds on a
broad diet that encompasses pretty much anything it
comes across, including figs, termites, leaves, fruits, seeds
and honey. Although most of its 'meat' comes from
insects and other invertebrates, it will take smaller
vertebrate prey that includes turtles, birds and small
mammals.[7] Its small size and dietary preference combine
to make this a species that would not regard us as prey
but, unlike the other species so far, the sun bear is
implicated in a number of injurious attacks on people.
These are not especially well documented, but one study
in 2013 did examine human–sun bear conflict in north-
east India.[8] Informal interviews in 40 villages in Mizoram
revealed 33 human casualties between 2000 and 2010,
with almost 80 per cent of the victims being male. Facial
injuries were common (reminiscent of the injuries
received by victims of hyena attacks), as were injuries to
arms, hands and legs (likely defensive injuries). There
were no fatalities and the incidents are described as
'accidental due to sudden encounters when villagers
ventured into the forests'. Sun bear attacks may occur, but
there is no suggestion of a predatory motivation.

Sun bear range extends into Northeast India, but it is
another species of bear that can properly be thought of

as 'India's bear'. The sloth bear is a mid-sized species with long, sickle-like claws on the front paws that are used to rip apart ant and termite nests. A specialist on these social insects, the sloth bear's mouth is well adapted to eating them. Long lower lips and a lack of upper incisors (the small biting teeth at the front of the mouth) allow sloth bears to suck up large numbers of insects at a time. Sloth bears are yet another species that wouldn't regard us as prey, but they are known to be aggressive in defence. Their temperament, a distribution that often puts them in close proximity to people, and their long, sharp claws mean that sloth bear attacks on humans are not unusual events, and can be serious. There are multiple studies of sloth bear attacks across India, but these are often conducted in a region-by-region or state-specific way. This makes collating and assessing attacks at national level tricky, as we have also seen with tigers (see Chapter Three) and leopards (see Chapter Seven), but even without that overview the situation is clear: sloth bears attack people and attacks can result in fatalities. Of 51 attacks that were the focus of a 2018 study in Central India, 34 resulted in serious injury and seven were fatal.[9]

There are local differences in sloth bear attacks and – as with other animal attacks – there are patterns of 'local outbreaks', with some regions having more attacks than others. One such area, the North Bilaspur forest division in Madhya Pradesh, has historically had especially high incidences of human attacks. A study of interactions between 1989 and 1994, showed that sloth bears accounted for a staggering 735 of the 1,094 human causalities from large mammals in this area (the rest were 138 leopard, 121 tiger, 34 elephant, 29 wild boar, 21 gaur – massive wild cattle, like a bull on steroids, 13 wolf and three striped hyena). Of those attacked by sloth bears, 48 were killed.[10] So, sloth bears may not view us as prey, but they are still a large animal that presents a very real threat to people living alongside them. As the authors of the study in Madhya Pradesh grimly summarise, 'Without active management [of human–bear conflict], local inhabitants will not support conserving the sloth bear.'

The black bear

To find bears that would consider us to be prey we need to switch continents and travel to North America. Here, we find three species of bear (two of which have a wider distribution than just North America) that will attack people, in some cases with predatory intent. The first of those is the American black bear. A very familiar species, this bear is found across Canada, the United States and into northern Mexico, although it is far less widespread than it once was (of course). These bears are mid-sized, dark-coloured animals, and although most individuals are black there is considerable variation across their range. Colour forms include cinnamon, light brown, dark chocolate-brown and silver-grey. We come across our old friend the subspecies once again, with 16 black bear subspecies often identified, including the glacier bear of south-east Alaska (a silver-grey bear), the Kermode or spirit bears of British Columbia (some of which have cream-coloured or white coats) and the cinnamon bears present across much of the northern parts of the species' range. Recent work examining the genetics and gene flow of black bear populations is inconsistent with these traditional subspecies, 'calling into question how to order taxa below the species level [*i.e.* subspecies]'.[11] Let's just say it's a work in progress.

Black bears are highly omnivorous, with a diet featuring berries, insects (especially ants), honey, seeds, nuts, salmon and other fish, and occasionally larger vertebrate prey. Black bears are known to take young deer when the opportunity arises and there are credible records of them preying on adult white-tailed deer.[12] A white-tailed deer is approximately human size, which should give us pause for thought, but there is evidence that black bears can take even larger prey. A report from Canada details a black bear kill site where the prey was an adult female moose.[13] A female moose is 400kg (882lb) of muscle. If a predator can tackle a moose, then a person is well within its prey spectrum.

In a study published in 2011, Stephen Herrero and colleagues collated records of black bear attacks between 1900 and 2009 by wild (*i.e.* non-captive) American black bears.[14] As with cougar attacks, the numbers are relatively low when compared with the level of attacks found with other predators in parts of the developing world. Nonetheless, at least 63 people were killed during the period, a frequency of one fatal attack every 21 months or so. Given the wide distribution of black bears and their relatively high population, these numbers do not suggest that black bear fatalities should be of particularly great concern, although of course each is a tragedy for those involved, their family, friends and communities. As with other predators we have met, the pattern across the species' range was uneven. Attacks in Canada, for example, were 3.5 times higher than in the United States despite bears being only 1.75 more common and human–bear interaction being much less likely. There were many regions with high bear numbers and no fatal attacks. Five states had estimates of more than 20,000 bears and no fatalities.

As is the case with cougars (see Chapter Seven), there is an interesting pattern in bear attacks over time. Of the fatal attacks recorded, 86 per cent occurred between 1960 and 2009, with a pronounced increase over the past three decades, especially apparent in the early 2000s. Once again it looks as if our increasing appetite for outdoor recreation may be a key factor in driving increased conflict with predators. It is also notable that attacks are most likely to occur in August (17 attacks), during which month humans are more likely to be active outdoors and bears are busy eating in preparation for hibernation.

Finding out more about attacks is important if we want to reduce – and hopefully eliminate – fatal human–wildlife conflict. The most common month for attacks may be August, but overall 85 per cent of attacks took place between May and September, with only seven attacks taking place after that, and just two attacks in April. Most attacks happened during daylight hours (70 per cent), with most of the remaining attacks taking place during the evening. However, 12 per cent of attacks occurred at night, darkness surely making a terrifying experience even worse. It doesn't get any better (and let's be clear, being attacked by a bear is already likely to be someone's worst experience) when we consider that bear attacks are often prolonged. In the eight attacks for which a duration could be recorded, seven lasted for between 10 minutes and several hours. Only one incident lasted for less than 2 minutes. Bear attacks are typically on single people (69 per cent of attacks), but some attacks are on groups of two or more and, rarely, these attacks may result in more than one fatality. Bears don't seem to have a preference for attacking men or women, or for any particular age group. However, analysing the number of attacks between the sexes and across different ages would need to allow for the number of such people likely to be in bear country, which is unlikely to be even across ages or, potentially, between sexes. The best conclusion is that 'any activity that brings people and black bear into possible close proximity may very rarely be associated with fatal attack'.[14]

I was at pains to point out that for other bear species, like the sloth bear, attacks may occur but they are not predatory. For the black bear the pattern is different. To classify an attack as predatory, Herrero and colleagues used a very strict definition: 'Predation: a series of behaviors, including searching, following or testing, attacking (capturing), killing, sometimes dragging a person, sometimes burying them, and often feeding upon them (to be classified as predation in our study, killing and some related behaviors had to be reported); vocalizing and stress behaviors by the bear were usually absent.' According to this definition, in 88 per cent of incidents the bear was judged to have acted as a predator. That is high. Nine in ten of the incidents they recorded were black bears attacking people as prey. Black bears will certainly attack us in defence, but that was not the case for most of the attacks Herrero and colleagues detail.

Male black bears are much more likely to be the perpetrators of predatory attacks; 92 per cent of attacks were carried out by adult or sub-adult (weaned but not yet breeding) males. When females attacked, in all cases the females had young with them. When it could be identified, in some cases (32 per cent) the bear was apparently injured, underweight or in poor condition, but in 68 per cent of cases there was no underlying health problem. The overwhelming conclusion again is that a great many black bear attacks are motivated by predation. Having a false notion that black bears will only attack if cornered, that they don't 'want' to attack us, doesn't seem to be especially true and isn't a useful mindset to have if you are in areas where bears live. Do all bears want to eat us all the time? Of course not. But given that a predatory motive was assigned to 88 per cent of attacks, then it seems reasonable to consider this as a part of the normal ecology and behaviour of at least some of the species. It is beholden on people to take this into account when hiking, cycling, camping or working in areas where black bears are present. I guess what I'm trying to say is, take it seriously.

The Asian black bear is a close relative of the American black bear. Sometimes called the moon bear or the white-chested bear (because of the pale crescent-moon marking on its chest), the Asian black bear has a wide range across eastern Asia. As we can expect, it is far less widely distributed than it once was and its current range is becoming highly fragmented. The species is an opportunistic omnivore, and will take small-to-medium prey including deer and wild boar, which puts it in the frame as a possible predator on humans. Certainly, attacks occur throughout its range, although studies indicate that these attacks are rarely fatal. A three-and-a-half-year study of victims treated at a care centre in Kashmir reported 120 Asian black bear attacks (and 35 leopard attacks) during the period.[15] Eighty-six males (adults) and 34 females (including two children) were attacked by bears during the period but only 1.6 per cent of attacks were fatal, compared with 48.5 per cent of leopard attacks. The far lower fatality associated with bear attacks than with leopard attacks is notable and it is interesting to compare the injuries caused by the two species. As the authors remark, 'Injuries of the cervical spine, cervical cord, major vessels of neck, pharynx and eye were striking observations in leopard attacks. Fractures of the upper limb, facial and skull bones were common in the victims of bear attacks.' Leopards are going for kill-strikes, which they deliver without much warning. Bear attacks on the other hand seem less directed and the injuries sustained are often defensive in nature.

Further study in this region confirms the idea that bear attacks are likely not predatory. Extending the cohort to 254 cases over 54 months, Asian black bears repeat a pattern consistent with self-defence rather than predation.[16] This doesn't reduce the effects of the attacks for the great majority of victims who survive, though. Bear attacks can cause considerable trauma both physically and psychologically. Proper treatment of such cases requires many sub-specialisms of medicine over a sustained period. In reality, this level of care is unlikely to be available to the victims of such attacks, which inevitably tend to occur in remote, rural locations. The authors of the Kashmir studies conclude that, 'Wild animal-inflicted injuries are a neglected part of trauma' and it is hard to disagree. Overall, human predation by bears is relatively rare, but international media coverage tends to be greatly skewed, as with cougar attacks, to North America. But when it comes to human suffering as a consequence of bear attacks, once again it appears to be Asia, and especially India, that should be in the spotlight. However, while Asian black bear attacks are likely not motivated by predation, it was clear that at least some (and possibly most, depending on location) American black bear attacks are. Heading back to North America (and far beyond) finds our last two, closely related, bear species, both of which are implicated in human predatory attacks – the brown bear and the polar bear.

Brown bears
The brown bear is a magnificent animal. It is a big bear, although length and weight vary greatly across its range. Bigger bears, found in resource-rich coastal areas, rival the size of the largest bear species, the polar bear. Brown bears found in the Kodiak Archipelago of south-west Alaska are the biggest of all. The Alaska Department of Fish and Game states that large males can weigh up to 680kg (1,500lb), which is a massive animal any way you want to look at it (my favoured way would be from a good distance, through binoculars).[17]

The brown bear is very much the quintessential bear and it is the species I tend to think of first when I consider bears. Even its scientific name speaks to its 'bearness'. *Ursus arctos* translates to 'bear bear' in Latin (*ursus*) and Greek (*arktos*). Found across Asia and Europe, in France, Spain, Italy, Greece, Nepal, China, Japan and a wealth of countries in between, brown bear populations are increasingly fragmented. However, the brown bear is still doing relatively well in North America, where it is more commonly called the grizzly bear. Mention of a widespread geographical distribution should by now ring some alarm bells labelled 'subspecies', and if you thought leopards and black bears were confusing then buckle up. In 1918, the American zoologist Clinton Hart Merriam proposed 86 subspecies just in North America.[18] Based mostly on small differences in skull measurements, Merriam's approach was that of a determined 'splitter'. In contrast, more recent analyses have tended to favour 'lumpers', bringing together groups to make fewer subspecies. The current situation is still complicated, though nowhere near Merriam levels of complexity. Some say five subspecies, others say nine and some stretch out to 15, but as with other species we have encountered the animals themselves are still very much recognisable for what they are, although in the case of the brown bear there is a further complication that can make then a little less recognisable.

Brown bears and polar bears, which we will consider shortly, are closely related. In fact, polar bears seem to have branched off from brown bears very recently, at least in geological terms. Evidence from a jawbone found

on the far western edge of the Svalbard archipelago in
Norway suggests that the point of divergence was
somewhere between 130,000 and 150,000 years ago[19],
while a genetic analysis suggests that the divergence
occurred between 479,000 and 434,000 years ago.[20]
Doubtlessly, the date will be further finessed (the two
species are the subject of intensive research), but all
indications are of a relatively recent evolution of polar
bears from a brown bear ancestor. Such a recent origin
points towards a strong similarity between the species
and that extends beyond mere zoological curiosity;
hybrids between the two species can occur. A spate of
hybrid bear sightings in the Canadian Arctic were
studied in 2017. These were found to be the result of a
single female polar bear that had mated with two male
grizzly bears producing hybrid offspring and offspring
that had 'back crossed' with grizzly bears. Climate
change is implicated in these and other hybridisation
events, with changes inducing overlap in the species'
range. Further genetic analyses have revealed extensive
genetic mixing between the two species at times in the
past that also correspond with periods of climate change
likely to have induced more mixing between the species.[21]

Brown bears, across their range, are incredible dietary
generalists and are often considered to have not just the
widest dietary spectrum of any bear, but one of the
widest of any animal. They take omnivory to extremes,
tackling pretty much everything from insects to salmon,
nuts to leaves and rodents to honey, but it is their
predation on large mammals that is of most interest.
This is a complex picture across the species. There are
differences between populations because the availability

of prey species will differ and there are differences between individuals. Overall, though, there is plenty of evidence to support the idea that brown bears can, and will, take larger prey including moose and reindeer calves, and livestock, especially sheep. They will also take elk and other large prey if the opportunity to take a weakened individual presents itself.[22]

Given their size, widespread distribution and proven ability to take down large prey, even if brown bears aren't dedicated predators they clearly tick all the boxes for a potential human-eater. Indeed, being larger and more geographically widespread than the American black bear, a species that does regard us at least at times as prey, would seem to make the brown bear a prime candidate. As ever, though, things aren't always straightforward in the world of human–wildlife conflict.

In 2019, a team of researchers led by Italian scientist Giulia Bombieri published a paper entitled 'Brown bear attacks on humans: a worldwide perspective'.[23] The team investigated 664 attacks on people between 2000 and 2015 covering most of the species' current range. Europe dominated the figures, claiming 291 attacks, versus 190 in the eastern part of their range and 183 in North America. This is interesting, since there are probably around four times as many bears living in North America than in Europe. Of course, what counts is the opportunity for bears and humans to come into contact, along with a host of other factors including livestock husbandry practices, hunting pressures and so on. By taking a broad overview of attacks across the world, Bombieri and colleagues were able to start to piece together the patterns and make some broad conclusions.

Most attack victims (99 per cent) were adults and of these 88 per cent were male. Half the victims were engaged in leisure activities that included hiking, camping, fishing, jogging and searching for mushrooms, while 28 per cent were working outside (logging, farming and so on) and the remainder (22 per cent) were hunting. This activity pattern probably explains the absence of children, for once, in attack statistics. Twenty-seven of the 123 attacks on hunters were on men hunting bears, which shows a certain karmic balance at least. For our purposes, though, it is the motivation for the attacks that is of most interest. Here, the team's conclusion was very different from that drawn by Herrero's team for American black bears. Nearly half of attacks followed an encounter with a female bear with cubs and a fifth were provoked by a sudden, surprise, encounter. Other trigger factors were the presence of a dog (17 per cent) and a bear attacking after being trapped or shot (10 per cent). Only 5 per cent of attacks were deemed to be predatory.[24]

By better understanding the nature of the attacks – and drawing out potential patterns of human and bear behaviour that result in negative encounters – we are better able to reduce the potential of such things happening. Of course, the very best way to avoid being attacked by a bear is not to be in an area where bears live. This type of advice is largely pointless and could even be offensive when applied to people living alongside most of the predators we have met. However, for many of the relatively very few people being attacked by bears in North America at least, the activity that caused them to be in bear country was recreational rather than existential, especially in recent decades. A bear attack may be terrible for the people involved, but it can also go badly for the bear. Headlines like 'Bear shot dead in France after attacking 70-year-old man', 'Grizzly bear shot dead after dragging woman from tent and killing her' and 'Grizzly bear killed after attack near Yellowstone National Park' demonstrate our reaction to bear attacks.

In Chapters Two, Three and Four, I supported the idea of killing predators after human attacks because in many places this can have the twin benefits of preventing future predatory attacks on humans and reducing extremely harmful retaliatory killing of multiple animals. There is, though, a very big difference between killing an animal that is seeking us out as prey and killing an animal that was simply defending its young, was taken by surprise or was being hunted. The killing of a brown bear that attacked a hiker is not the same as the killing of a lion that has killed and eaten a farmer in Tanzania or a crocodile that has taken a fisherman in Indonesia. The freedom to explore the 'wild' for leisure comes with responsibilities to the species you may encounter and that define the very wilderness you seek out.

As always, the authors of the brown bear study point out, quite rightly, that brown bear attacks generally are rare events. Nonetheless, they occur and in fact account for the majority of attacks in regions where brown bears and black bears coexist. Another study of bear attacks, this time in Alaska between 1880 and 2015 by Tom Smith and Stephen Herrero (who led the team focusing on black bear attacks earlier) showed that 88 per cent of bear attacks there were by brown bears, with black bears accounting for just 11 per cent of attacks.[25] The mathematically astute among you may have noticed that 1 per cent of attacks in the Alaska study aren't accounted for by brown bears or black bears. That tiny proportion of attacks is by the final species in this survey of bears – the polar bear.

Polar bears

The polar bear is the largest of all bear species, and indeed is the largest land carnivore, with very big males weighing up to 700kg (1,543lb). These are animals that spend much of their life around the edges of the sea ice of the frozen Arctic, where both human and bear densities are low. The other bears we've met are mostly omnivores, with a surprising dietary preference for insects like ants or for plants like bamboo. Such exotic choices are firmly off the menu for polar bears, though, who are severely limited in their dietary options by the extreme environment in which they live. As a consequence, polar bears are described as a 'hypercarnivore', a species whose diet is more than 70 per cent meat. Polar bears are, if anything, hyper-hypercarnivores since, although they will eat berries, kelp and other plant material, to a first approximation their diet is seals. Ringed seals and bearded seals make up most of their diet, with the skin and especially the blubber being exactly the sort of calorie-dense meal a large, warm-blooded mammal needs at low temperatures. Of more interest when it comes to risk of human predation, they will also hunt and kill larger land animals, including musk ox (shaggy, cow-like beasts that look as though they've been transported to Earth from another planet) and reindeer, known as caribou in North America.

Polar bears have a reputation of being 'stalkers and killers of man', as Smith and Herrero put it. But, as they also point out, this reputation is not well founded. Indeed, polar bear attacks are so rare that Smith and Herrero were unable to draw many conclusions or discuss any useful patterns in the data from attacks in Alaska. James Wilder of the US Fish and Wildlife Service led an international team of researchers in a study of polar bear attacks from across their range, and this study gave the scientists more data to consider, although not that much more. From 1870 to 2014 they documented 73 attacks in Canada, Greenland, Norway, Russia and the US (Alaska). These attacks resulted in 20 fatalities and 63 people injured. Some attacks could not be confirmed, but of the 63 that could, 59 per cent were predatory. Of the bears that attacked, 61 per cent were in poor condition, 'skinny or thin', and 65 per cent of bears that attacked for predatory reasons were in below-average condition. Overall, 88 per cent of fatal attacks were predatory in nature and 83 per cent of fatal attacks involved consumption of the victim.[26] So, polar bears will kill and eat us, but they do so rarely and the behaviour seems largely to be driven by bears being in poor condition.

What to do in bear country

One of the most common questions regarding bear attacks is what to do if such an attack happens. Rarely have I come across a question with so many conflicting answers. Taking a consensus of advice, the best strategy is to play dead, while making yourself look simultaneously big, small, threatening and not-threatening. This is a tough combination to pull off. Playing dead is advice often followed by the darkly humourless observation that it will be good practice for what follows, but it would be good to have something more concrete to work with; gallows humour is not especially useful in keeping you alive. Here is better advice, given by the US National Park Service in an article helpfully entitled 'What should you do if you encounter a bear' on the BBC's News website.[27] Step one is to recognise the signs of a bear not being best pleased. This seems relatively simple; clacking teeth and huffing or sticking out the lips are all good signs that you are too close. Another sign you are too close to a bear is – and it's hard to put this more plainly – you are close to a bear. You may accidentally end up near a bear, but never plan to be in that situation. If you can tell a bear is clacking its teeth, then maybe things are already going south...

The second useful piece of advice holds for most predators – don't run. Ever. You won't outrun a bear or anything else with teeth and claws, and you'll probably trigger a chase response that won't end well. So, however much you want to run – and you really will want to run – don't. Stand your ground. And ground means ground; don't be tempted to climb a tree, because bears are good tree climbers. If you think you have problems with a bear on the ground then you have even more problems 10m (33ft) up a tree with a bear for company. At least on the ground gravity isn't adding to your woes. This advice is of course irrelevant for polar bears, because the nearest tree may be hundreds of kilometres away.

If a bear charges at you then remember this; most charges are mock charges. This will not be a hugely comforting piece of trivia if the situation arises, but it remains a fact. If a brown bear charges then stand your ground (see above) and deploy your anti-bear pepper spray, which you will be carrying because you are in bear country and you aren't an idiot. Bear spray works. Smith and Herrero report a 93.3 per cent success rate of bear spray 'altering aggressive behaviour', one assumes for a better outcome for people. This is a better rate than a rifle, which achieved 75 per cent success. The reason firearms are less effective than you might think is really very simple; real life isn't what you see in movies. Bear attacks can happen fast and from close range. About half of all encounters with bears in the Smith and Herrero study occurred within 10m (33ft), with a strong element of surprise for both species. There is very little time to react. Unless you are very practised, have a suitable firearm ready to shoot (*i.e.* a cartridge of sufficient calibre to kill a bear chambered, with the safety off), and are willing and able to shoot a bear in a way that will immediately incapacitate it without thinking, then the chances of even getting a shot off are incredibly low in many encounters. In reality, you are likely to be paralysed with fear or indecision for a second or two at least. And, of course, you shouldn't be carrying a loaded rifle with the safety off; you are far more likely to have an accidental discharge and kill someone else, or be killed by someone else, than you are to encounter a charging bear. For this reason, some people opt to carry a large-calibre revolver, which is largely idiot-proof to use and doesn't have safeties to fumble with. But if you've been brought up on

Westerns then you have entirely the wrong idea about how quickly a handgun can be withdrawn from a holster, and how easy it is to hit a target quickly and accurately under extreme pressure. You aren't Clint Eastwood; carry bear spray.

If the bear does convert a mock charge into an attack then you have two options, if you have the wherewithal to do anything at all; you can play dead or you can fight back. If it is a defensive attack then playing dead is the right thing to do. Drop to the ground, curl up into a ball, soft belly region inwards, and stay still. If you have a rucksack on your back then it will help to protect you, as will your bunched-up arms and legs. You're going to get hurt though, and quite possibly very badly. A defensive bear will keep its head low and its ears back according to Kerry Gunther, Yellowstone's bear management leader. So, be sure to be taking careful observations during the second or so of extreme panic as it closes in. A predatory attack is different. A predatory bear will keep its head up and ears erect. In that case, you need to fight back. Of course, the other way to tell if an attack is predatory, is that after a few minutes your strategy of playing dead isn't working. I'm sure you can see the problem with this, so really do try to observe the bear closely as it charges.

I don't know about you, but I don't fancy the odds against a bear regardless of its motivation. Many people do survive bear attacks, but I don't really want to be part of that cohort. I would far rather avoid an attack in the first place, and once again the US National Park Service provides useful advice. The number one thing you can do is to follow what it terms 'viewing etiquette'.[28] This code of practice is aimed at national park visitors who may encounter bears, but it is applicable to viewing most wildlife safely and respectfully. If you want to observe the predators covered in this book, then following these rules would not be a bad plan.

1. Respect the animal's space. Use binoculars, keep your distance and if the guide you are with insists on going closer (as many do, seeking tips from their guests), politely but firmly tell them that you are close enough. I can't stress this enough; poor outcomes for people and wildlife nearly always involve people being too close. Buy binoculars; keep your distance. Many TV naturalists behave appallingly around wildlife, so don't look to them for an example. If you only do one thing, then respect animals' space.

2. Linked to point one, never crowd an animal. If there is a big crowd gathered around any wildlife sighting (and this is horribly common when viewing African and South Asian predators in popular tourist destinations) then don't be part of it.

3. Stay in groups and minimise noise. Not only will you be safer, but you'll probably see more wildlife. That said, animals don't like being surprised, so if you are in dense forest or areas with low visibility then talk quietly among yourselves. This will allow animals to hear you and get out of your way, which is what most species want to do most of the time.

4. Stay on designated trails. This should be a no-brainer, but stick to paths and roads and

follow signs whether you are on foot or in a vehicle. If signs say 'Stay in your vehicle', then stay in your vehicle. It isn't difficult.

5. Leave 'orphaned' or sick animals alone. You certainly aren't Dr Dolittle and you very likely aren't a qualified wildlife vet. Wildlife doesn't like being handled and mothers don't like their offspring being handled. If you think something should be done then report it to whatever authorities there are, but don't intervene.

6. Leave your dog at home. Dogs bark, are associated with bear attacks, carry diseases and disturb ground-nesting birds. And leopards eat them.

7. Don't feed wildlife; remember the mantra 'a fed bear is a dead bear' and insert any other species name you want. If you feed animals and they become used to people, then they will become a problem. Quick story: I stayed at a place in South Africa that had lots of baboons running around. They were cute, especially the younger ones, and tourists were feeding them despite notices and admonishments to the contrary. There were some big males in the group and they were getting very bold. One jumped on to a dining table, scattering the group of diners and causing general mayhem. When I went back six months later there were no baboons running around. They had all been shot. If you were one of those people feeding them because you thought they were cute or 'felt sorry for them' (as one tourist told me) then you were responsible for the

action that had to be taken. So, don't feed wildlife.

Remember, ultimately you are responsible for your safety and the safety of wildlife. That is a responsibility that you need to take very seriously if you have chosen to put yourself into the privileged position of being able to experience wildlife first hand.

CHAPTER NINE

Canids

Many people likely to be reading this book may never see any of the predators we've covered in the wild. In terms of negative interactions with predators, it is highly unlikely that many of you will think that you've ever had one. And yet I pretty much guarantee you have.

I started back in Chapter One with the example of dogs attacking people in the UK to illustrate the point that attacks by predators aren't necessarily predation. Of course, an interaction doesn't need to go as far as an attack, with injuries, to be considered negative. I would say most people have had some kind of negative interaction with a dog, perhaps many more than one, although such interactions only rarely result in bites. That said, in a study of 1,280 households in Cheshire, a quarter of respondents reported that they had been bitten at some point in their lives.[1] A quick straw poll of friends suggests that this proportion, though seemingly high on first look, is probably in the right ballpark. The conclusion back in Chapter One, though, was that dog attacks in the UK cannot be considered predatory and nothing in the Cheshire study suggests otherwise. While this is also the case for the vast majority of dog attacks and bites received globally, it is not universally so. Man's best friend can eat us – and not just via scavenging in the sad cases where owners have died and their pets are left with no choice.[2]

In 2017, it was widely reported that a man in central Russia's Khanty-Mansi region was attacked and killed by a pack of stray dogs. Police reported finding the 'remains of the body' the next day and many sources led with a headline that involved the word 'eaten' somewhere. In a parallel to cases in previous chapters, including lions and army ants, alcohol was involved. The attack was picked up by the world's media because much of it was caught on CCTV and the footage clearly showed a drunk man staggering around.[3] It is debatable whether a sober man would have fared much better once an attack was in progress, but sober people tend to make better decisions and avoid incidents in the first place. In this case, the victim had been feeding the stray dogs for two years, but on the fateful day had tried to get past them without food.

When we say 'dog' we are normally talking about the domesticated dog in all its many forms. The term refers biologically to a much wider group of species that are more technically referred to as canids. For our purpose, most of the diversity of the canids can be dealt with quickly, since most members of the group are not potential predators of humans. Ultimately of course we are heading towards wolves, but there is a surprising amount of diversity to work through before we get there – and some species that can cause us more than a few issues, including direct predation.

The 'foxes'

There are a number of smaller canids that are referred to by the word 'fox', including the bat-eared fox, the Arctic fox, the fennec fox and the red fox. Some of the 'fox' species group together into the 'true foxes' and different 'foxes' are sprinkled around in other groups, but regardless of where they are placed they aren't looking upon us as prey. The red fox is the largest of all of them and it is a 10kg (22lb) animal when fully grown, although large males might make 15kg (33lb). When compared to a dog of similar size, foxes are much slighter. These are lightweight, small animals that live on their own. A fox isn't going to take you down as prey any more than a corgi (which is about the same weight) would.

South America has an interesting group of canids that evolved separately from the rest of the group, which includes the crab-eating fox and the stubby tailed, stocky little bush dog. It also includes the maned wolf, which actually looks like a fox and yet is neither a true fox nor a wolf. The maned wolf is the largest of the South American canids at 20–30kg (44–66lb) and, while not a large predator, it is reaching a size that could be a problem for a smaller adult or a child. However, the maned wolf seems universally to be described as timid and even its livestock predation is mostly limited to poultry. Threatened by habitat loss and fragmentation, and by deaths on roads, the maned wolf is not a predator of humans.

Another charismatic canid with some severe conservation pressures is the African wild dog, also called the African painted dog, the cape hunting dog and, more recently, the painted wolf. These highly marked, highly carnivorous dogs are well built, reaching 20–25kg (44–55lb) in eastern Africa and more than 30kg (66lb) in southern Africa. They are highly social, living in stable groups that can number more than 20, although typically comprise somewhere between four and ten individuals. They are undeniably handsome animals and their tight social structure and range of group behaviours make them one of the more popular species to see in Africa, as well as the subject of a number of notable recent TV documentaries. However, this modern-day popularity has not saved them from the usual trident of predator conservation pressures. Habitat loss, prey depletion and direct persecution have seen the species exterminated across much of its historical range and the current population, hard as it is to estimate, is thought to be about 6,000–7,000 individuals.

Active persecution has occurred largely as a consequence of the perceived threat wild dogs pose to livestock. Certainly, wild dogs can and do take livestock, but – as always – the pattern may be complex. In a study in northern Kenya, for example, the number of livestock taken by wild dogs was exceedingly low, at one attack per 1,000km^2 (386 square miles) per year, but only where wild prey remained, even if at low densities.[4] Where wild prey had been seriously depleted, livestock attacks ballooned, causing considerable local economic harm despite the overall regional cost being low. We have seen this pattern before several times; the costs of predators may be highly localised and those effects can be overlooked if we only consider the bigger picture. While patterns of livestock predation may be complex and locally variable, one thing seems to be universal: African wild dogs are not considered to be especially dangerous to people. Indeed, it is a commonly repeated statement that there are no confirmed reports of African wild dogs attacking – let alone eating – people. This isn't true, although, in fairness to African wild dogs, the incident I'm about to replay was highly unusual.

In 2012, two-year-old Maddox Derkosh fell over a railing at Pittsburgh Zoo and Aquarium separating visitors from a pack of African wild dogs held in an enclosure, described as a 'dog pit', below the railing. Maddox fell just under 4m (13ft) into the enclosure, where 11 dogs rapidly attacked, killing the toddler in front of his horrified parents.[5] Some of the dogs were able to be called off, but one especially aggressive individual had to be shot. Maddox bled to death from massive injuries and there is no reason to think that he would not have been eaten were the dogs given the chance. It is hard to fathom the depths of grief and horror Maddox's parents faced that day – and continue to face – but the incident is a strong reminder that many predators view children differently to adults. While there are no reports of African wild dogs attacking or predating people, the Pittsburgh Zoo incident makes it clear that, at least when it comes to a child, they could do so with ease. But, for now at least, they are off the list of species that predate us.

A canid that we cannot take off the list of possible human predators is the Australian dingo. The dingo is a ludicrously controversial species. Everything about it is disputed. Some say it's a domestic dog, while others have asserted it to be a subspecies of either the domestic dog or the wolf. Then there are those who have pronounced it to be a full species. The date when it arrived in Australia is disputed, its relationship to other dogs is disputed, its ownership and status as a pet is disputed, its ecological role as a predator is disputed and its impact on livestock is disputed. Whether a dingo took a nine-week-old baby girl called Azaria Chamberlain from a tent in 1980 was also heavily disputed, in a long-running case that featured in a film starring Meryl Streep as Azari's mother Lindy (who is attributed with the cry 'A dingo ate my baby'). Azaria's body was never found and Lindy was tried for murder, convicted and sentenced to life imprisonment in 1982. The conviction was overturned in 1988 and a third inquest in 1995 returned an open verdict. The case continued, and in 2012 it was ruled that Azaria has indeed been taken and killed by a dingo. Overall, dingo attacks are very rare, but when they occur children are often the victims. For example, a spate of attacks, mostly involving young children, occurred in 2019 on Fraser Island off the coast of Queensland.[6] In an echo of the Chamberlain case, this included a toddler being rescued from a dingo's jaws after the boy had been dragged out of a campervan. It seems reasonable to assume that such attacks are motivated by predation and, while an adult is probably safe, there is a very small but real chance of young children potentially being identified by dingoes as prey.

One canid we can definitely dismiss as a human predator is the dhole, a beautiful medium-sized canid of central, south and South East Asia. Slightly larger than a red fox, the dhole is a sociable animal that lives in clans of 5–12 individuals, with groups of 30 or even 40 sometimes reported. These clans can break up into smaller groups at certain times of the year, then reform later, with a more fluid structure than we see in other

social canids like African wild dogs. Being mob-handed
can compensate for smaller size (and army ants are the
ultimate example of that), but despite strength of
numbers, dholes are not reported to have attacked
people. That said, their favoured prey size, animals in
the 30–40kg (66–88lb) or higher range, suggests that
they could. The fact that dholes have been reported to
take on tigers, putting them at bay up trees and even
killing them adds some further weight to the idea that
dholes *could* take on a person. This potential perhaps
underpins the fear revealed by a study of people's
perceptions of dholes in southeastern Thailand.[7] The
study showed that dholes were perceived as potentially
dangerous, with half of respondents believing they will
attack a person despite, as this study confirms, the fact
that there has never been a report of a dhole attacking a
person anywhere. It is understandable that a group-
living predator, which can take on a tiger, might be
perceived as dangerous, but an incorrect negative
perception against a predator can be just the fuel
needed to drive negative outcomes, including direct
persecution.

Despite some negative perceptions, the Thailand
study also reports a large contingency of villagers who
held positive attitudes towards dholes. This allows for
some optimism, but leveraging these attitudes into
positive conservation outcomes requires education. The
authors of that study make proposals for education that
fall into two approaches. Broadly, these approaches
could apply to predator education elsewhere and with
other species. The first approach can be thought of as
'immediate and local'. An issue for dholes is confusion

with golden jackals (which we'll look at next). The tendency to lump all canids together as 'forest dogs' leads to any negative interactions occasioned by one species to be transferred to all species. This 'tarring with the same brush' could well extend to include predators in general, not just those that are closely related or similar. The authors of the Thailand study conclude that 'the first step to cultivating a positive social climate for dholes in Thailand is to teach people the physical, behavioural and ecological-niche differences between dholes and jackals. We highly recommend the development of education materials that pictorially differentiate dholes and Asiatic jackals.' Such an approach is simple in principle, but success requires buy-in from communities, access to communities and resources to develop appropriate materials, none of which are necessarily straightforward in practice. Other 'immediate and local' interventions could include encouraging people to gain personal experience with dholes. By watching them, learning about them and, crucially, learning how to behave appropriately around them, fear could be reduced.[8]

The second approach towards education is broader, putting the focus more on structural changes to wider education. School curricula can include more ecological education that incorporates predator interactions and ecology in ways that highlight their role in the wider ecosystem. We increasingly understand education as a lifelong venture, and it makes sense that predator education should follow that model and continue beyond school. Training people working in protected areas in communication and outreach, for example, and providing carefully assembled interpretation material for people living in areas with predators, can all help to reduce fear. Finally, and a point raised by the authors of the Thailand study, people could be encouraged to 'experience nature'.

The next canids we can touch on lightly are jackals. Side-striped and black-backed jackals are African species that are closely related and broadly similar. Both species are medium sized, weighing in at around 8–12kg (18–26lb), with the side-striped jackal generally being slightly larger. Neither are especially fearsome in appearance and their dietary preferences, which include invertebrates, rodents and scavenging on carcasses, do not immediately suggest an animal that will cause us any particular concern. However, they will prey on young antelope and smaller species such as dik-diks, and this aspect of their ecology can put them in conflict with livestock farmers, who fear that jackals may take their stock. This fear is not unfounded; jackals will take sheep, goats and poultry. During the lambing season especially, jackals can be a problem for sheep farmers and this has led to considerable persecution across their range. Shot, trapped, coursed with dogs and poisoned, jackals nonetheless remain relatively common and widespread. Indeed, in parts of South Africa, changes in land use and other factors have led to something of a resurgence in jackals and this has inevitably led to more conflict. However, neither African jackal species has been implicated in predatory attacks on humans and so need not concern us further.

The other species known as a jackal is the golden jackal. More closely related to the wolf and domestic dog (all three species are in the genus *Canis*) than it is to the African jackals, the golden jackal is found from Eastern Europe through south-west and southern Asia, as well as parts of South East Asia. More wolf-like in appearance than the African jackals, golden jackals are still not especially large, an average-sized individual being around the same size as a larger side-striped jackal. Like their distant African cousins, golden jackals are implicated in livestock predation, which has led to extensive persecution. However, while African jackals are not especially aggressive towards people, there are reports of golden jackals attacking people across their range. For example, the *Times of Israel* detailed 'a spate of jackals attacking humans in northern Israel' in February 2021,[9] while in November of that year in India a single jackal in the Katihar district was reported to have attacked 40 people, with eight suffering serious wounds.[10] The Israel attacks were suggested to have occurred because people were feeding stray cats and leaving garbage around, both of which lured jackals into built-up areas where they could lose any fear of people. Rabies was also mentioned as a possible factor. A viral disease that causes inflammation of the brain, rabies can present with a variety of symptoms that can include increased aggression. Rabies is a common factor in animal attacks and seems likely to have be a factor in the India incident. What doesn't seem to have been a factor in either Israel or India was predation. However, there is at least one credible report of golden jackals killing, and eating, a person.

In June 2019, the *Telegraph India* ran a story with the self-explanatory headline 'Jackals kill and "feed on" 9-year-old boy in Murshidabad'.[11] Murshidabad is a city in West Bengal in east India, and the attack in question took place in the village of Gopalnagar. Zeeshan Sheikh had been picking grass for goats when he was set upon by a group of seven golden jackals. A youth named Rafiq saw Zeeshan's body being dragged by jackals and described the horrific incident to the *Telegraph*: 'I ran screaming with a hammer, but it was too late. I am surprised that we did not hear him [Zeeshan] scream. By the time I reached the spot, his abdomen and throat had been ripped apart. Half-eaten flesh and internal organs were lying around.' A graphic description of what was undoubtedly a predatory attack.

Jackals are actually widely known to attack people in India, a country that sees more than its fair share of human–wildlife conflict. One study, reported in the *Telegraph*, documented 220 attacks between 1998 and 2002 in the central state of Chhattisgarh, although none was fatal. Anindita Bhadra, a canine behaviour specialist at the Indian Institute of Science Education and Research in Kolkata, commenting to the *Telegraph* on the attack on Zeehsan, asked, 'Is jackal behaviour changing in eastern India? This could be a possible subject for future research.' It is a certainly a valid question, especially given the size of the group that attacked Zeeshan. Only around one in five golden jackal groups exceeds three individuals so a group of seven (assuming that the group size is correct) would be highly unusual. At the moment, though, all we can say for sure is that at least one person has been killed in a predatory attack by golden jackals.

Golden jackals bring us a step closer to the ultimate potential canid human predator, the wolf, but before we get to that most problematic of dogs we have a couple more species to consider in the genus *Canis*, the first of which is the coyote. A kind of bargain-basement wolf, the coyote is a medium-sized canid that in many ways is an American equivalent of the golden jackal. Similar in size and habits, but more closely resembling a small wolf visually, the coyote is very widely persecuted across its range, primarily because of its habit of killing livestock. While we can debate the rights and wrongs of the situation endlessly (and many people do), the fact is that coyotes do kill large numbers of smaller livestock, especially sheep. As a consequence, they are routinely shot, poisoned, trapped and otherwise killed, sometimes in exchange for bounties. While the killing of coyotes is fuelled by livestock losses, coyotes will also attack people and in some cases those attacks are predatory. A study published in 2009 analysed 142 coyote attacks that resulted in 159 victims.[12] The authors defined 'attack' robustly (in my opinion) as being an incident that resulted in a bite – in other words a serious incident that resulted in physical injury. This removed incidents that may have been distressing (quite possibly to both species), but would be hard to characterise in terms of behaviour or motivation. Attacks were defined as being predatory in nature if the coyote(s) 'directly and aggressively pursued and bit a victim, causing multiple or serious injuries (often to the head and/or neck)', or where the coyote 'bit the victim in the head or neck and attempted to drag them away'. Of the attacks they documented, 37 per cent were judged to be predatory, and predatory attacks were significantly more common on children. In 7 per cent of attacks the coyote was confirmed to be rabid and none of the attacks was fatal. Of course, the usual caveats apply here and the chances of being attacked by a coyote, regardless of motivation, are very small. Nonetheless, we have to include coyotes as predators that, at least at times, consider us to be prey.

The other wolves

All of this canid diversity and potential (rarely met) for human predation is of course simply build-up for the main event – the wolf. Emblematic of the 'wild', yet deeply feared, wolves are ingrained in culture and folklore in many parts of the world, echoing their wide distribution, at least historically. They have variable ecology and differing human relationships across their range, which makes generalisations difficult, and a dizzying array of subspecies that folds in further confusion. Nothing is easy when it comes to wolves, even defining what we mean by a 'wolf'.

When most people think about wolves they are probably thinking of the grey wolf *Canis lupus*. This is the animal of *Little Red Riding Hood*, which must be kept from the door and hides in sheep's clothing. However, within the genus *Canis* are two other species that also carry the glory and burden of being termed 'wolves': the African wolf and the Ethiopian wolf. The African wolf, or the African golden wolf as it is also known, looks very much like the golden jackal. Indeed, for a long time the African wolf was considered to be a subspecies of golden jackal. Its genetics are a fascinating tangle of canid ancestry, and its classification a confusing journey into the world of species and subspecies, but in terms of human predation things are thankfully simpler. There seems to be no good evidence that African wolves see us as a potential meal. The other species carrying the name 'wolf', the Ethiopian wolf, is even less of a threat to us. One of the world's rarest canids, this beautiful animal is red on its sides and back, with a white belly and chest, and a long and narrow head. It has an unusually restricted diet for a canid, feeding almost exclusively on small rodents. To the all-too-familiar threats to predators we can add canine distemper and rabies, diseases the Ethiopian wolf can catch from domestic dogs. If ever there was a species under serious and imminent threat, it is the Ethiopian wolf.

The grey wolf

Let's start with the basics. The grey wolf is a large canid, with a variable coat that can be grey, almost black, brown, mottled or even white. It is the largest of the non-domesticated canids and only just misses out on being the largest of any canid to a few exceptionally large domestic breeds. By large, we mean large. The heaviest wolf ever recorded, at least according to Guinness World Records, was an adult male from Yukon, Canada, that weighed 103kg (227lb). This individual was exceptional, for sure, but males weighing in the 60–80kg (132–176lb) range are not uncommon in some populations.

'Some populations' is a crucial qualifier because the grey wolf has a very wide geographical range and as a consequence has a large number of identifiable

populations. It has been extirpated from many parts of
its historical range (including the UK and much of
Europe and the US), but still occupies an impressive
sweep of territory that spans much of the northern third
of the world. In what is by no means an exhaustive list,
grey wolves can be found in Canada and the United
States, in Spain, France, Italy, the Balkans Poland and
much of eastern Europe, parts of Scandinavia,
Greenland, Russia, much of central Asia, Turkey, Israel,
Jordan, Saudi Arabia, Iran, Iraq, India, and eastwards to
China and Korea. Wolves exhibit considerable variation
across that range and between populations. Nearly 40
subspecies have been designated for *Canis lupus* at one
time or another (some of which are now extinct) and
there is constant disagreement over how wolves should
be classified. A good example of this concerns wolves in
the United States. There, two forms of the grey wolf, the
red wolf and the eastern wolf, have variously been
considered to be full species, subspecies, hybrids
between grey wolves and coyotes, or some combination
of these. How these animals are classified has important
implications for conservation. If we impose protective
measures on the grey wolf then does that cover the red
wolf if it is designated to be a different species? The
temperature is already high when it comes to wolves and
conservation, and confusion over species classifications
hardly cools things down. The situation is perhaps best
summed up by the introductory sentence to a recent
paper on North American wolf taxonomy: 'The
evolutionary origins and hybridization patterns of *Canis*
species in North America have been hotly debated for
the past 30 years.'[13] The same is true of other subspecies.

The variation in subspecies is well illustrated by differences in size between subspecies. The 103kg (227lb) Yukon wolf was probably a member of a subspecies known as the northwestern wolf (also known as the Canadian timber wolf, the Alaskan timber wolf and the Mackenzie Valley wolf) but may have been a member of the Interior Alaskan, or Yukon, wolf subspecies. These two subspecies are the largest and second largest grey wolves. Adult males of both subspecies have a very good chance of weighing more than 50kg (110lb) and, as we've seen, have the potential to be considerably heavier than that. The smallest subspecies of wolf, on the other hand, is the Arabian wolf, native to the Arabian Peninsula, the Sinai and Negev deserts of Egypt and Israel, and Jordan. An average adult male Arabian wolf weighs just over 20kg (44lb). That is about 40 per cent of the weight of an average northwestern wolf and less than 20 per cent the size of the 103kg (227lb) record-holder. This is a tremendous size variation for a species and size is important when it comes to assessing the risk posed by predators to humans. I have some friends with an English bull mastiff that weighs 101kg (223lb). Rocco is a truly massive dog, and 2kg (4.4lb) lighter than the Yukon wolf. On the other hand, a 20kg (44lb) wolf is lighter than a fully grown Labrador. Given the choice, I'd rather not be facing down any wolf, but a 20kg individual is clearly much less of a potential threat than one exceeding 100kg.

Size is only one factor in assessing the potential threat posed by a predator. Group behaviour is also important because grouping up can assist greatly in hunting prey that may be challenging or impossible for an individual to tackle. Wolves are famously pack animals, but even their social structure can vary greatly across their range. Wolf packs are described that range from two to 20 individuals, with some subspecies (like the Arabian wolf) having typically smaller pack sizes than others (such as the European wolf). Of course, there is also the 'lone wolf', a single individual living outside of a pack structure. These individual wolves are often younger animals (perhaps between one and two years old), dispersing from the pack into which they were born. In many wolf populations, lone wolves are a minority and dispersing animals will join up with a pack, but in some cases lone wolves may be more common, especially if larger prey is locally unavailable. As with everything to do with wolves, it's complicated.

With at least some of the basics in place, we can now ask the question: do wolves consider us prey? The simple answer to that question is that yes, there are good examples of wolves killing and eating people, but far fewer cases than you might expect. In 2002, a report entitled 'The fear of wolves: a review of wolf attacks on humans' was published by a large international team of authors led by John Linnell.[14] The report deals with wolf attacks during the past few hundred years, collating records from across the world. It remains a definitive source for human–wolf interactions, although as we will see there have been some important changes in wolf distribution in Europe since it was published.

The report makes clear distinctions between different types of attack. Attacks by rabid wolves are distinguished from defensive attacks where wolves have been cornered or otherwise provoked. As discussed previously, rabies can present with different symptoms and these can progress in different phases. One phase, characterised by hyper-reactivity to stimuli and a pronounced tendency to bite, is termed (for obvious reasons) 'furious rabies'. It seems that wolves develop especially severe 'furious rabies' and can bite a large number of people in a single episode (as many as 30 in one case). The report concludes that the majority of attacks recorded were by wolves with rabies. This conclusion is based on a preponderance of historical data; rabies is far less of a problem in many parts of the world than it was although such attacks still occur in the Middle East and Asia.

The report also details unprovoked predatory attacks, which are the attacks of primary interest in considering whether wolves view us as prey. It turns out that despite widespread fear of wolves, and their infusion into mythology and culture as beasts to be feared, such attacks are incredibly uncommon. Mostly predatory wolf attacks centre on Europe and parts of Asia, and the best recorded examples are from pre-twentieth-century France, Estonia and northern Italy. Together, these three regions account for several hundred people being killed by wolves between around 1750 and 1900. This number is bolstered greatly by a series of attacks between 1764 and 1767 in the Gévaudan area of France, where records indicate that more than 100 people were killed by wolves. However, outside of these regions – and beyond the 'Gévaudan episode' – predatory attacks are rare. Other pre-twentieth-century records from Scandinavia have far lower numbers. Norway has but a single record, a six-year-old girl killed in 1800, while Sweden has four children killed between 1727 and 1763. Eleven children and a woman were killed in 1820–1821 in Gysinge, central Sweden, by what was believed to have been a single wolf that had escaped from a life of captivity where it would have become habituated to people. Finland has a number of attack episodes, mostly occurring within five clusters through the 1800s. Around 62–75 children were killed during these episodes and just two adults. When wolves are taking humans as prey they are almost exclusively targeting children.

Attacks during the twentieth century in Europe become even less common, but are also strongly focused on children. Attacks tend to occur episodically, suggestive of a single wolf or pack becoming habituated to people, then developing a penchant for predating children. Five children were killed in Poland in 1937, four in Spain between 1957 and 1974, and 36 around Kirov, Russia, between 1944 and 1953 in three separate episodes. In 1944 and 1950, 22 children were killed in the 'Kirov episode', in 1951 and 1953 four children were killed in the 'Oritji episode' and a further 10 were killed in the 'Vladimir episode' in 1945 and 1947.[15] These episodes have been disputed, and the background to these disputes is worth a detour, because it reveals much about the factions that emerge, even within the scientific literature, when it comes to the risks posed by wolves to humans.

There are probably more wolves in Russia than in any other country, and they are widely distributed across this vast and geographically variable area. Wolves in Russia have been subject to sometimes intensive control efforts motivated by real and perceived human–wolf conflicts, including livestock killing and the potential for human predation. In 1982, a book entitled *The Wolf* by Mikhail Pavlov was published. Pavlov was a hunter and game manager who viewed wolves through glasses that were very far from rose-tinted. He saw wolves as vermin and the authors of the attack report liken the tone of his work to that of a 'personal crusade on his [Pavlov's] part to "tell the truth about wolves" *i.e.* that they are dangerous to humans'. The report concludes that Pavlov was not 'an objective and unbiased observer'. The reason why Pavlov's motivations and biases are relevant is that he presented in *The Wolf* details of attacks that many researchers and conservationists had a hard time believing. It is usually difficult to believe events and facts that go against your thinking, but when they are presented by someone who is clearly 'in the other camp' it can be nearly impossible.

The 'Kirov episodes' were almost unprecedented and, given Pavlov's clear aversion to wolves, were interpreted by some as an attempt to whip up anti-wolf feeling. Balancing such accusations is the fact that Pavlov supplied names, locations and details for the Kirov and Oritji episodes, lending some credibility to his depiction of events. But there is another factor that lends weight to Pavlov's claims – a set of unusual circumstances occasioned by the Second World War.

The Kirov and the Vladimir episodes together resulted in 32 dead children between 1944 and 1950. The Second World War was just ending at the start of these episodes and wolf control was hardly high on the list of priorities. Men were fighting wars and firearms were unavailable, so wolf-control programmes had been suspended during the war and did not resume during the post-war recovery period. This inevitably led to a higher population of wolves, but also to wolves that had not experienced hunting pressures. Echoing Rajeev Mathew's comments on tigers in India, this would probably have led to an increasing population of predators no longer conditioned to be fearful of humans. Pavlov also reports that, at the same time, wild prey population levels were low, putting pressure on an expanding wolf population to find food. These predator/prey factors are joined by human social factors. The war and post-war period were incredibly tough years for many in Russia and – just as in Bardiya National Park, Nepal – people were probably forced into closer contact with predators in order to find vital resources. The balance of evidence overall suggests that, despite his underlying biases, Pavlov's accounts of human predation by wolves in post-war Russia are credible.[16]

Wolves in North America
In the decades leading up to 2002, predatory wolf attacks were extraordinarily rare, despite wolves becoming more abundant during this period in many key parts of their range. They were so rare, in fact, that there are no documented predatory attacks in North America during the entire twentieth century, and only two attacks that may have been predatory in the twenty-first century. The second, and clearest, example of a predatory attack was on Candice Berner, a 32-year-old teacher, who was attacked by a group of wolves while out jogging in Chignik, Alaska, in 2010. She died of 'multiple injuries due to animal mauling' and a subsequent investigation ruled out rabies as a contributory factor. The report

concludes that the attack was predatory in nature although the evidence seems more to lean towards an opportunistic predation event than a premeditated stalk. The forensic evidence at the scene allowed for some partial reconstruction of what may have occurred and it seems quite possible that the 'encounter may have come as a surprise to both parties'.[17] Berner was alone, small in stature and running, none of which are useful attributes in a sudden encounter with a predator. There is evidence that she reversed her direction, indicting that she may have tried to run away from the wolves when she became aware of them, releasing a predatory response from them. Although there was no strong evidence that the wolves were stalking Berner, they nonetheless killed her and had begun to eat her body. It was clearly predation.

Another probable wolf attack occurred five years earlier in 2005. Kenton Carnegie, a 22-year-old geological engineering student, died after being attacked while walking near Points North Landing in the Canadian state of Saskatchewan. This attack proved more difficult to piece together. Some features of the attack suggested that a bear was responsible, but other features pointed to wolves. The fact that wolves were known to be in the area, that bears had not been seen around and that wolf tracks were present seemed initially to point very strongly towards wolves. However, further investigation only managed to conclude that Carnegie had been killed by a 'predatory animal'. The best interpretation of events seems to be that he was killed by wolves and that is now the generally accepted story. If so, then his death marks the first predatory attack by wolves in modern times in North America.[18]

Child lifting

The recent deaths in North America do not follow the very clear pattern established in Europe because neither victim was a child. However, with such a low sample size it is hazardous to draw any firm conclusions. What we can say for sure is that predation focusing on children is the norm for other locations. In India, for example, records from three states in the last few decades of the twentieth century show at least 200 children being killed in the Hazaribagh region of Bihar state between 1980 and 1995; 50 children being killed in just eight months of 1996 in the eastern region of Uttar Pradesh; and nine children killed in six months in 1980 – 1981 in the Anantapur region of Andhra Pradesh state.[19] The taking of children, especially at night and from dwellings, is known in India as 'child lifting' and the clustered, episodic nature of the killings is linked to specific individuals or packs of wolves.

Wolves that overcome fear of humans or habitation, or that have not developed such a fear, can find easy prey in children, but it should be stressed that such behaviour is highly unusual in more recent times. I asked Rajeev Mathew, who you'll remember keeps records of predatory attacks on people in India, about wolves in modern-day India. He told me that 'attacks by wolves are largely a thing of the past now and we have very few of them these days. I do not have any records for this year [as of 30 November 2021], but one last year where many were injured in a single pack attack. Most of these attacks happen because the wolves are rabid. Except for lifting children, wolves do not by and large attack humans. Child lifting by wolves has become a thing of the past – the last major attacks on children were in the 1990s in Andhra Pradesh and Uttar Pradesh [the attacks referred to above].'

Problems ahead?

The Linnell *et al.* wolf attack report sums up the situation succinctly when it says that 'in those extremely rare cases where wolves have killed people, most attacks have been by rabid wolves, predatory attacks are aimed at children, attacks in general are unusual but episodic and humans are not part of their normal prey'. This is good news, especially when we consider the fact that wolf populations are increasing in many places and – especially in parts of Europe – we are also seeing considerable expansion of wolf range. This is great, but we should also be cautious to ensure that conservation successes don't backfire.

People do not universally view the expansion of range and increase in abundance of wolves (or other predators for that matter) as a good thing. Entirely rational fears of livestock losses often do the heavy lifting in arguments against predators, but close behind them comes fear for personal safety. Even if misplaced, or exaggerated, people's natural fear of being attacked and potentially eaten by a predator needs to be taken very seriously and treated respectfully. These concerns are the primary stumbling blocks for the predator-focused rewilding initiatives that dominate recent media narratives around conservation. Central to many people's vision of a 'rewilded wilderness' (whatever that might mean in practice) is the restoration of extirpated predators. The prospect of wolves roaming the Highlands of Scotland may excite you (it certainly excites me), but it does not excite everyone. It is against this background of natural range expansion and potential reintroductions that we need to consider the three factors other than rabies identified by the Linnell *et al.* report as being associated with attacks on humans. All three are likely to be exacerbated by increasing wolf numbers and expanding range, regardless of the mechanism by which that is achieved.

We met the first factor, habituation, in Chapter Three, when Mathew spoke of tigers losing their fear of people. We saw this effect again when we considered wolves in post-war Russia and the Kirov episodes. To a certain extent it may also play a role in attacks by urbanised hyenas and leopards. Simply put, if predators get used to us, and no longer fear us, then there is an increased risk of attacks. What this means is that by protecting predators to conserve them we also inevitably lay the ground for habituation and conflict. Predators are also potential tourist attractions and having more people in the landscape could further exacerbate the issues of habitation. Tourism is often touted as being a major benefit of rewilding – and it is hard to disagree – but we will need to be very mindful indeed of our pervasive and potentially harmful influence on the animals we seek out.

The second factor is more obviously a problem: provocation. The examples Linnell *et al.* give are clear-cut – trying to kill a trapped or cornered wolf, or entering a den with pups. The reality, though, is that without detailed knowledge and experience of being around predators much of what we do could be provocation. When predators increase in number, encounters with people with limited experience or knowledge become far more likely. Mix in a habituated predator – and a natural sense of curiosity in both parties – and a negative situation can rapidly develop. In Thailand, we saw the argument for educating people about how to interact with dholes and a similar argument holds for human–wildlife interactions anywhere. If you don't know how to act around predators, you have no business being around predators. And watching every David Attenborough and Steve Backshall documentary doesn't give you what you need sadly. In fact, it may well set you up to be part of the issue. If that sounds harsh then it is only because I've seen too many people getting out of vehicles when they shouldn't, getting too close when they didn't need to, and generally being a problem for human and wildlife safety to have too much patience. Respect an animal's space and keep your distance.

The final factor Linnell *et al.* identify is arguably the most important when it comes to considering the risks associated with natural range expansion: highly modified environments. They conclude from their extensive analysis that the majority of predatory attacks have occurred in what they define as modified environments. Modification in this sense includes environments with little or no natural prey, the use of garbage and livestock as food by wolves, and children unattended or used as shepherds. They also list poverty (no surprises there) and limited access to weapons as being contributory factors in a list that encapsulates many of the features we might expect in heavily human-dominated landscapes. Disease outbreaks in India associated with leopard human-eating behaviour from Jim Corbett's time (see Chapter Seven) also fall under this category. When wolves move into such modified landscapes, attacks can occur. Linnell *et al.* point out that it is unlikely that there was so little prey available during periods of wolf-attack episodes that wolves had no choice but to prey on children. Rather, they suggest that the close association of wolves with people in landscapes with limited prey (made worse perhaps by scavenging on rubbish near settlements) make such events far more likely.

Living with wolves

The canids are a potentially fearsome group of animals, but despite ability, opportunity and motive very few represent any real threat to us in terms of predation. Rabies has a large part to play in unprovoked attacks, but credible reports of canids hunting, killing and eating us are remarkably rare. In the modern world, even the largest and most fearsome of all the canids, the wolf, is not a major threat despite a reputation that drips into our minds from an early age via nursey rhymes, stories and expressions. Many groups are hugely supportive of wolves and it is not hard to see why. They are magnificent animals. But we should be careful not to get carried away with our enthusiasm and forget that conservation, especially of predators, is complex. Human fears, whether real or perceived, are vital to understand and accommodate if we are to have landscapes with predators present and free from persecution. Wolves have steadily spread through Europe over the past decades and have now been spotted in the Netherlands, Belgium and Luxembourg. It is wonderful to see these animals expanding and doing well, but this will inevitably push them up against human populations that are unused to – and perhaps unhappy about – living with wolves. Local livestock practices will need to change, perhaps drastically, and people may need to alter their behaviour in some areas to accommodate wolves. But this can be done, if we chose to do it.

The people of Sanabia-La Carballeda in northwestern Spain provide a model of how local practices can change to function in a landscape with predators.[20] The area has one of the densest wolf populations in western Europe, but local shepherds have refined their husbandry to such an extent that sheep losses are 'a trickle'. They have packs of guard dogs that stay with the sheep and, in the same way that livestock is protected from lions, the sheep are locked up at night. Of course, such practices come with a cost, and few people like change, but if changes can be made then predators can provide benefits for local communities. For example, in Spain, up until September 2021 when it was banned, it was legal to hunt some wolves and the local council gained revenue by selling hunting rights. Wolf tourism also provides some income, but this doesn't always trickle down to those directly affected by wolf presence. The simple fact is, no matter what benefits there may be, when you live with predators someone has to pay the price. It might be feeding an extra guard dog, building a *boma* or a sheep hut, employing extra shepherds or losing some livestock to predation. Or, in the worst case, the price may be paid in human lives. But, whether it is wolves in Spain or lions in Tanzania, if we want predators to thrive we must look after the people who pay the price.

CHAPTER TEN

Fish, Lizards and Primates

Predators, by definition, kill and eat prey. As I've shown over the past nine chapters, there are numerous species that can make us and our livestock prey in the right circumstances – and some can do so frequently. This can cause considerable problems for the people affected and, through pre-emptive and retaliatory measures, the predators themselves. Consequences of human–predator conflict add to ongoing near-global issues of habitat loss and prey depletion to drive down predator numbers in the ways that have become horribly familiar over the course of the past nine chapters. Before considering what we can all do, even if we aren't on the front lines, to help predator conservation, there are some other species that warrant attention. These are species that have been known to kill and eat us, but not frequently enough to deserve a full chapter all to themselves.

Fishy tales
I said in Chapter One that I wasn't going to talk about sharks because so much has already been written about 'SHARK ATTACKS!', but since I am going to be discussing a range of other species that attack and eat us then they probably do need some consideration. (Note that I'm going to consider sharks overall, rather than by species.) In 2021, the Florida Museum of Natural

History's International Shark Attack File records 137
'alleged shark–human interactions'. These are negative
interactions (because no one maintains databases of
positive wildlife interactions), and investigations
revealed 73 unprovoked and 39 provoked shark bites
worldwide that year. Those incidents resulted in 11
shark-related fatalities. Across the world, 64 per cent of
the bites were in the US and 60 per cent of those were in
Florida; as with saltwater crocodiles (see Chapter Four),
Australia might have the reputation, but the real action
is elsewhere. These numbers are low when compared
with some of the predatory interactions described in
previous chapters, but there is another feature of shark
attacks that makes them stand out from, say, those of
crocodiles or tigers; what people were doing at the time
of the attack. Unlike the attacks of many other species,
we have excellent knowledge of people's pre-attack
activities: 51 per cent were surfing or taking part in
some other board sport; 39 per cent were wading or
swimming; 4 per cent were snorkelling or freediving;
and 6 per cent were engaging in what the Shark Attack
File describes as 'body surfing or horseplay'.[1] No one
was subsistence fishing, collecting water, washing
clothes, bathing or going to the toilet (although this last
activity is an assumption on my part and I guess there
could be some multitasking). Overall, shark attacks
seldom happen and rarely result in fatalities; most bites
occur out of curiosity or as a consequence of mistaken
identity. Most importantly, the people who are attacked
didn't have to be there. For many predators, relative
poverty is a very strong factor in determining who is
attacked. That is simply not the case for sharks. You

choose to surf or swim, but you don't choose to get
water from the only river around.

The oceans, deep, dark and mysterious, hold a morbid
fascination for many and the nightmare of being attacked
and eaten by some 'sea monster' is a common trope in
seafaring tales and legends. That fear is real but, while
the ocean is clearly a dangerous place, the predators that
dwell there pose very little threat to us overall. The far
less-feared freshwater 'monster fish' don't get the same
attention but can nonetheless attack and kill people,
although the numbers are very low. There is only a small
number of freshwater fish species around the world that
are large enough, with a diet carnivorous enough, to
consider a person as prey. The goonch catfish, also called
the giant devil catfish, is one such species, found in rivers
around the Indian subcontinent. Goonch catfish can
reach 2m (6ft 7in) or more in length and have long been
subject to suspicions of human predation, enhanced by
observations of them feeding on human remains from
funeral pyres in the Kali River.[2] If we take those
observations, together with reports that linked
drownings and disappearances to suspected catfish
attacks, then we have to accept at least a tentative
conclusion that such attacks are possible and in all
likelihood have occurred. Another bulky catfish with a
reputation for attacking and eating people is the wels
catfish. Potentially reaching 3m (9ft 10in) and 200kg
(441lb), the wels catfish is found across central, southern
and eastern Europe. As is the case with goonch catfish,
there are some stories of human predation, including
credible historical accounts of catfish being found with
the remains of children in their stomachs and tales of

fishermen being taken. However, given the very wide distribution of these fish in water courses extensively used by people in countries with modern record keeping and media, the scarcity of credible accounts suggests that these fish are not routinely seeing humans as prey.

As we saw in Chapter Five with army ants, a lack of size can be compensated for by strength of numbers. The freshwater fish with the fiercest reputation for human predation, the piranha, is a good potential example of the mob principle in action. The word 'piranha' can be used to refer to any of a multitude of species present in South America and especially the Amazon basin. The diversity and relationships of these species is still being untangled, but for the purposes of potential human-eating it is the red-bellied piranha that is the species most clearly in the frame. Red-bellied piranha can reach 50cm (1ft 8in) and 4kg (8.8lb), but most individuals are around 30–35cm (1ft–1ft 2in) long. Not a small fish, but certainly not a large one. Their reputation, though, is as a ferocious shoaling predator, which descends in great numbers on hapless prey in the water, 'skeletonising' them in seconds in a boiling mix of water, countless razor-toothed fish and blood. It is a terrifying prospect and one supported by some elements of truth. These fish do sport incredibly fierce-looking teeth, but ecologically they are far better described as timid scavengers rather than ferocious predators. They can, and do, take down smaller mammals and waterfowl at times,[3] but there are very few credibly evidenced accounts of them attacking and eating people. Those accounts we do have often record people being eaten after they had died from heart failure and drowning.[4] There are, however, a very small number of reports of people being killed by them in predatory attacks. These include a drunk man (now there's a common theme) who jumped into a river in Bolivia and died from piranha bites in 2011.[5] Police suspected suicide in this case, which is an interesting angle to add to human–predator relationships. In 2015, a six-year-old girl, Adrila Muniz, died in Brazil after her grandparents' canoe capsized and she fell into the water. Piranhas ate most of the flesh from her legs and she died at the scene in the sort of horrific incident many would associate with piranhas, but is mercifully rare.[6] Piranhas do sometimes bite people swimming in rivers, but these attacks are generally single bites and are associated with the fish defending their brood. These fish, despite the stories, don't form predatory shoals that roam around looking for prey; in fact, they group up to keep safe and avoid becoming prey themselves.[7] They can and have attacked people, but these attacks are rare and seldom result in fatalities. However, I'll be honest and say that the images of Muniz after her attack are enough to put me off swimming in water where they might be present.

Here be dragons
Crocodilians and, to a lesser extent, some large snake species carried the banner for reptiles back in Chapters Four and Five, but there is another predatory reptile species that can cause us problems. The Komodo dragon is the largest lizard in the world, reaching 3m (9ft 11in) and 70kg (154lb). Its size alone puts it firmly into the 'possible human predator' category and

the fact that it can kill pigs, goats, horses and even water buffalo does nothing to lower that status. Endemic to a few Indonesian islands (notably Komodo), the Komodo dragon is categorised as endangered and has suffered (surprise, surprise) from range contraction and population decrease in recent years. The species has been known to kill people, and it seems reasonable to assume a predatory motivation in at least some cases, but these animals are not major players when it comes to human predation. In 2007, the death of an Indonesian boy (unnamed, obviously) after an attack by a Komodo dragon was widely reported in the world's media.[8] The attack was described as a 'mauling' and the boy died 30 minutes afterwards from blood loss. The incident happened in the dry season, when natural prey is scarce, so the attack could have predatory, but regardless of motivation it was exceptionally unusual. The boy's death was the first human death from a Komodo dragon for 33 years. Since these animals roam around the residential area of the national park, where they live in reasonable numbers, it seems fair to say that we are not generally featuring on their menu.

Looking in the mirror

The final two species that warrant a brief mention are, surprisingly, two primates. The first is a weird species with a range of complex behaviours that can be hard to fathom at times: humans. Human cannibalism is a well-recorded, but incredibly complex phenomenon that is often bound up with deep cultural and spiritual significance. Predation, the hunting and killing of a person expressly for the purpose of eating them, is very different from cannibalism related to grieving, post-battle rituals or desperate situations where human flesh is the only food available. That said, 'famine cannibalism' (the eating of people during times of low food availability) and 'survival cannibalism' (where people eat others in extreme survival situations) have a long and grim history. Uruguayan Air Force Flight 571 famously crashed high in the Andes mountains in October 1972. The subsequent 72-day survival story, at the end of which 16 people were rescued, involved the difficult, well-considered and well-documented decision to eat the remains of those who had not survived.

While eating famine and accident victims is technically scavenging, it would be a very fine line indeed between scavenging and predation if someone were 'helped on their way'. That fine line has doubtlessly been crossed countless times. A very famous example of line crossing involves Tom Dudley and Edwin Stephens, who were shipwrecked when the yacht *Mignonette* broke up in a gale in the south Atlantic on 5 July 1884. Two other men, Edmund Brooks and Richard Parker, were also on board and the four managed to escape to a lifeboat before the *Mignonette* went down. Long story short – Parker, the inexperienced 17-year-old cabin boy, got ill on 20 July, probably from drinking seawater, and fell into a coma. Four or five days after that, Stephens held Parker down and Dudley stabbed him in the neck, killing him with the remaining men (mostly Dudley and Brooks) going on to consume Parker's body. The survivors were rescued, and the two men involved in the killing were initially sentenced to death for Parker's killing. Eventually this was commuted to a six-month prison term after a trial that became famous for setting the legal precedent that necessity (in this case, the killing of someone to eat who was certain to die anyway) is not a defence for murder.[9]

The second primate species, and the last species on our roll-call of minor or unusual players in human predation, is one we might not suspect of having predatory intentions. Chimpanzees are, like us, a member of the group of primates known as great apes. Chimp babies are very cute, but that is not a description usually applied to fully grown chimpanzees. A male chimp might stand more than 1.5m (4ft 11in) high and weigh 70kg (154lb) or more. That size is already concerning given the experience of other species in earlier chapters, but far more worrying if you were ever to confront one is chimps' power. Chimps are very strong. It is common to read accounts of them being five times stronger than humans. The eminent primatologist Frans de Waal, talking to the magazine *Scientific American,* states that chimps have strength that is 'utterly incomprehensible ... five times the arm strength [of] a human male'.[10] Such claims seem pretty incredible given our similar build and musculature, and others have cast doubt on chimp 'super-strength'. In 2017, a study was published that concluded chimps were 'only' about 50 per cent stronger than humans.[11] Regardless of the magnitude, the direction is clear and an adult chimp would probably have the upper hand in most physical encounters with a human. Despite this, wild chimp attacks on adult humans are not well recorded, although captive chimp attacks do occur.

Attacks on children, and especially very young children, however, are a different story. Chimps living in the Ruteete region bordering the Kibale National Park, Uganda, were recorded as having killed three babies and maiming several others in 2004. Attacks have been reported from the region before and the technique the chimps used in most cases was identical to the one they use to hunt red colobus monkeys – limbs are bitten or broken off and the prey disembowelled.[12] Other accounts tell a similar story. In 2014 a two-year-old, again in Uganda, was taken from his mother, Ntegeka Semata, who reported that the chimp 'broke off the arm, hurt him on the head, and opened the stomach and removed the kidneys'.[13] Frodo, a male chimp in the Gombe National Park, Tanzania, made famous by the work of Jane Goodall, took and partially ate a 14-month-old child. In a news piece relating the incident, Frodo's behaviour was described as 'part of the natural hunting behaviour of chimpanzees: it seems they can view human babies just as they view the young of other species such as colobus monkeys and baboons, as potential prey'.[14] The stark conclusion was that, 'This was not the first case of human babies being taken by chimps in the Gombe area.' Chimpanzees are a surprising but well-supported addition to the list of animals that can hunt and eat us.

CHAPTER ELEVEN

What can we do?

I want to live in a world with big predators. I am guessing you do too. I want my grandchildren, great-grandchildren and their great-great-great grandchildren, to live on a planet where tigers stalk, lions roar and crocodiles lurk. I often tell people how much they need to appreciate insects and the less glamourous animals, and that we need to realise that all biodiversity should be respected and admired, but to be honest I'm as excited about seeing a lion or a leopard as anyone. Probably more so. Sure, insects are massively important, but I have never got goosebumps spotting an aphid or a surge of excitement watching a butterfly. There is something very visceral, very primal, about big predators. A world without them roaming free is a terrible prospect.

My emotional response to predators is all very well, but there are also rational reasons why I want big predators around. Predators of all sizes play an incredibly important role in regulating ecological communities. Without them, the species they prey on can increase unregulated, causing all kinds of knock-on ecological effects. For example, the UK lacks large predators and partly as a consequence has an overabundance of deer. Deer strip bark from trees and eat the understorey, and if they are present in large numbers they can affect the way woodlands grow and the cover provided for nesting and feeding birds. This can be a major problem for species like nightingales, willow warblers and chiffchaffs. Predators have another important and counter-intuitive ecological role – they can increase biodiversity. Predators will often focus their hunting on the most abundant species and by taking them out provide ecological breathing space for other species. This means that predators can increase the number of species present overall by preventing any single prey species from becoming dominant. So, as well as the 'goosebump' reasons, maintaining predators in ecosystems makes sound ecological sense.

Maintaining predators through ongoing conservation efforts is one side of the coin, but the flipside is to actively introduce predators in to an area or to encourage natural range expansion. Range expansion has hardly been a common theme of the preceding chapters, but it has nonetheless appeared in the stories of several species' recovery, including cougars and wolves. Predator introduction on the other hand is becoming a dominant theme in some of the conservation narratives developing around 'rewilding'.

Lacking a coherent single definition, or much agreement over aims and methods, 'rewilding' broadly encompasses the notion of converting land back into some semblance of wilderness for biodiversity benefits. What 'wilderness' really means – and what level of management and human intervention are acceptable – are continuous and sometimes acrimonious points of contention. To be honest, rewilding has become a term thrown around so casually in some quarters that it runs the risk of losing any real value. When someone lets a roadside verge grow for a couple of months before mowing and calls it 'rewilding', or 'rewilds' their garden by digging a pond, then I think we're definitely entering a stage of definition-creep. However, some form of wilderness reclamation seems likely to be important if we are to navigate through the current crisis towards a future that can sustain us and maintain a high level of biodiversity. What land is used, how it would affect different stakeholders and what goals we should be heading towards are just some of the issues in the tricky waters that rewilders must navigate, but it is the active introduction of predators – such as wolves or lynx (as is being discussed in the UK) – that tends to feature most prominently in media discussions of rewilding.

Clearly, if we want ecosystems (rewilded or otherwise) to function without substantial human intervention then predators are crucial. But, equally, predators come with baggage; negative implications, both real and imagined, for our safety and that of our livestock. If we accept that we want a world where predators can thrive, then we must also accept that we want a world that will impose a burden on some people. Now, I am making an assumption about the location of my readership, but it is my guess that the people who will continue to carry most of that burden are very probably 'not us'. An overwhelming conclusion of many of the preceding chapters is that it is the rural poor of the developing world that are most affected by predators, despite some pockets of conflict in the developed world over wolf and cougar range expansion. It is a brutal reflection indeed of the inequality of the world that people are maimed, mutilated and killed by the same species that many of us pay to watch on our TV screens or, if we're lucky, to see in the wild on holiday.

So, what can we do to make the world a better place for predators – and for people? Most people can't hop on a plane to go and help directly and, bluntly, even if they could they probably lack the skills and knowledge required to be of any use. There is always plenty of enthusiasm for 'getting conservation done', which is great, but misdirected enthusiasm can be a problem. There is no better illustration of this than the ambitions of the group Rewriting Extinction. Launched in 2021, with a celebrity-laden video, Rewriting Extinction promised to focus on 'environmental solutions, and not all the problems'. Quite how you develop solutions without focusing on the problems is not explained, but TV presenter Chris Packham makes it clear in the closing statement that whatever has been done before (presumably all conservation up until that point) is 'messing around'. His last words hang in the air as a call to arms: 'Get it done.' But what is 'it'? The campaign generated considerable attention, as is virtually guaranteed when celebrities are involved, but from the conservation community on Twitter there was something of a scornful pile-on. For those actually getting 'it' done, the notion that Rewriting Extinction was going to achieve the aim implicit in its name through a bit of can-shaking and celebrity endorsement was laughable and insulting.

For those who decide to take a more a direct approach there is usually a steep and unpleasant learning curve. While driving around in an artfully muddied-up 4X4 poking a radio antenna out of a window looking for big cats might do well on Instagram, the #mood isn't the reality of conservation on the ground. Stress, lack of sleep, burnout, mental health issues, a constant lack of resources, deprived living conditions, lack of basic comfort, nagging constant fear, exhaustion, extremely low and unreliable pay, and an ever-growing level of cynicism don't play so well on social media, but they do tend to be the reality for many working in conservation in the wild places of the world. Luckily, though, there are some things anybody can do right now without spending a single penny, without the need for 4X4s or khaki trousers, and without even leaving the house. These things aren't so much about 'getting it done' as they are about getting in the right mindset to understand what 'it' is and what 'done' means. Here are my top five suggestions:

1. Stop thinking conservation is about animals. It isn't.
I recently had the huge pleasure of spending three days driving through Namibia in the company of a leading light of Namibian conservation, Maxi Louis. Louis works directly with communities in Namibia to try to balance some of the difficult issues I've explored in this book. We had some very long car journeys, more than one biblical deluge and plenty of time to talk. And talk we did. We spoke about communities, economics, colonialism, history, politics, apartheid, social mobility, enterprise, tourism, racism, sexism, the pandemic, nutrition, ecology, habitat, tribalism, linguistics, climate change and more besides. You know what we didn't talk about? Animals. In three days of sometimes intense conservation conversations wildlife hardly surfaced as a topic – and when it did it was nearly always because we'd seen something through the window.

What those conversations emphasise is that good conservation is, above all, about people. Most especially, it is about the communities that live alongside wildlife, on or around land that people far away (including me) pompously call 'habitat'. So, stop thinking that people are 'the problem', stop waggling accusatory fingers at 'too many people' in 'other countries' (while conveniently ignoring your own over-consumption) and start paying attention to those who actually live with wildlife. Until we realise that conservation is about people, we aren't going to 'save' anything in any meaningful way. Which leads me to point two...

2. Listen. To the right voices.
The modern world is a loud and brash place. If you want to be heard then you need to get access to the right media outlet, shout loudly and ideally get some celebrity shouters to back you up. None of that makes it easy for the right voices to be heard and so it is hardly surprising that we rarely hear the conservation voices we need to listen to. Instead of community members, overseas leaders, scientists and experts, we have actors and sports stars breathlessly telling us about animals 'on the brink of extinction' and how we should dig deep to save them 'before it's too late'. But have you ever had a celebrity ask you for money to buy an enclosure so that people can wash their clothes without being eaten by a crocodile? Or for money to pump-prime small enterprises to stop people from being converted into a tea towel full of meat and bone fragments by a tiger? No? Me neither. Solving actual problems is a lot less glamourous than bottle-feeding orphaned lions and asking for money to 'save the species'. We need to start listening to those on the

front line of conservation – and that means listening to communities, not comedians.

3. Include don't exclude, support don't impose.

I really can't say this enough, but we all have to stop thinking that conservation is something imposed on people 'over there' by organisations and people 'over here'. Conservation is riddled with 'white saviour complexes', and plagued by grandiose schemes that have zero basis in reality, and little or no connection with people and communities. It is quite possible that you've given money to organisations guilty of such approaches. Without a bit of critical thought, pretty soon you could find yourself on the slope towards supporting the exclusion of people from their lands, the devaluing of local knowledge and customs, the stripping of dignity from communities, and even violence and other abuses. These are not things most people want to endorse, but they are the very real potential consequences of imposed conservation and exclusionary agendas.

4. Accept complexity.

There is no single solution for predator conservation, or any conservation for that matter. What works for one species will likely not work for another, no matter how tightly we cross our fingers and hope. In fact, what works for one group of one species in one place might not even work for the same species 10km (6 miles) down the road. Or the same group three months later. It's a horrific web of nuance and complexity even before we take point one into account and adding people rarely made any situation simpler. When you then mix in the existential risk that predators pose, and all the baggage they bring with them, then the mess becomes even messier. We just have to accept that. We also need to accept that what constitutes an ideal solution might look very different depending on where you stand. If that place is many thousands of miles away, then perhaps you need to accept that your opinion weighs rather less than that of people standing much closer to the problem. Which leads me on to point five.

5. Have some empathy.
Do you have any idea what it is like to have your child's face literally taken away by a hyena? To have a loved one killed, gutted, half-eaten and dumped in the forest by a tiger? Or to have parts of your father's body cut out of a crocodile's gut on the riverside in the afternoon sun? I certainly don't and I sincerely hope you don't either, but around the world there are many people who do and many who have known far worse horrors. Each terrible incident ripples out through communities, with profound direct and indirect effects. Many people live in real fear of it happening to them or to those they love. Pretending that isn't the case, or underplaying the human cost of predation, is failing to acknowledge reality. Ignoring problems because they don't fit into our cosy preconceptions may make us feel better, but it isn't the way to find solutions.

★

Back at the start of Chapter Two I asked you to imagine you were a villager in Tanzania. My idea was to get you to think about what life might be like in a world where large predators roam. I don't think any of us can really imagine what it is truly like to be someone else, but we can try to listen, to educate ourselves and to think about what other people's lives might entail. Back in Chapter Two, that process came with the realisation that some people live very close indeed to nature. That intimacy with the natural world, something many of us seek out for relaxation and invigoration, can be especially dangerous if it involves 180kg (397lb) big cats or 5m (16ft 5in) crocodiles. Once we allow ourselves to develop some understanding of, and empathy with, those who live alongside biodiversity in such a meaningful and raw way, then we might just start to listen to what they have to say and learn from them. From there, perhaps we might begin to accept the complexity of their lives and start to support what – and who – really matters in conservation. That can only be a good thing, for people and predators alike.

Acknowledgements

I would like to thank Jim Martin of Bloomsbury for his support, Emily Kearns for her editing skills, and Borra Garson of DML Management for her advice and encouragement.

Special thanks go to the many people I have spoken to during the writing of this book, but especially Rajeev Mathew and Amy Dickman. I would also like to thank Rami Tzabar and everyone at the BBC Radio Science Unit for their unstinting support.

Finally, I want to thank Caroline Mills of the University of Gloucestershire, whose tireless and supportive management has allowed me to combine all of the things I love doing into something resembling a job.

References

Chapter 1: Introduction

1 The documentary *Big Game Theory* can be found at www.bbc.co.uk/programmes/b067x5w1

2 This report for DEFRA details victims and the dog breeds involved in fatal attacks, and is available through UK Parliament at http://data.parliament.uk/WrittenEvidence /CommitteeEvidence.svc/EvidenceDocument/ Environment,-Food-and-Rural-Affairs/Dangerous-Dogs -Breed-Specific-Legislation/written/83473.html

3 A Wikipedia page is devoted to fatal dog attacks in the UK and is regularly updated with links to press articles and other reports https://en.wikipedia.org/wiki/List_of_fatal _dog_attacks_in_the_United_Kingdom#2010%E2%80 %932019

4 Honey badgers and wolverines have a particular reputation for fierceness and strength. Both are members of the mammalian family Mustelidae, a diverse group that also includes otters, the European badger, stoats and weasels. The wolverine is the largest of the mustelids and has a reputation for ferocity and strength that needs little introduction to anyone familiar with *X-Men* in comic-book or film form. Despite their reputation, according to a study published in 2019, "there are no official records of a wolverine attacking a human" (www.tandfonline.com/ doi/abs/10.1080/1088937X.2019.1685020). Honey badgers, or ratels, are considerably smaller than a wolverine, but have a reputation for ferocity that, if anything, is greater. They will face down lions (sometimes more than one at a time) and leopards without fear. Reports that they charge humans and tear off testicles are widely repeated (including on the Botswana episode of the motoring series *Top Gear*), but seem not to be reliable. However, as an aside, a game ranger told me of a honey badger that was cornered in

a shower block by a young lioness. After an almighty amount of noise, a blood-soaked honey badger emerged and walked off into the bush. The lioness was dead.

5 The death of Kirsty Brown was widely reported at the time, *e.g.* www.nationalgeographic.com/news/2003/8/leopard -seal-kills-scientist-in-antarctica. Subsequently, a review was set up to examine the interactions between human and leopard seals. The report is available at www.bas.ac.uk /data/our-data/publication/interactions-between-humans -and-leopard-seals and concludes that 'the death of Kirsty Brown does show that leopard seals can display predatory behaviour towards humans'.

6 A longer version of this story can be read at http:// southafrica.co.za/legend-of-harry-wolhuter.html   and Wolhuter documents the attack in his book *Memories of a Game Ranger*, published in 1961. Wolhuter's knife looks like a wooden-handled carving knife. It wouldn't be your ideal first choice from a line-up of weapons with which to fight a lion.

Chapter Two: Lions

1 See Packer *et al.*'s 2005 paper, Lion attacks on humans in Tanzania, for a photograph of just such a hut, www.nature. com/articles/436927a.

2 The paper in 1. above and this 2010 paper by Kushnir *et al.* are sources for Tanzania lion attack numbers: Kushnir, H., Leitner, H., Ikanda, D. & Packer, C. 2010. Human and ecological risk factors for unprovoked lion attacks on humans in southeastern Tanzania. *Human Dimensions of Wildlife* 15(5):315–31.

3 www.tandfonline.com/doi/abs/10.1080 /10871200903510999

4 The programme can be heard at www.bbc.co.uk/sounds/ play/w3csz9dt

5 Packer, C., Swanson, A., Ikanda, D. & Kushnir, H. 2011. Fear of darkness, the full moon and the nocturnal ecology of African lions. *PloS one* 6(7):e22285, available at https://

journals.plos.org/plosone/article?id=10.1371/journal.pone.
0022285#s3

6 One of the many newspapers and online media outlets
to cover the story, the *Daily Mail* headline writer really
pulled out all the stops. Note the reference to partial
nudity, faeces and the use of capitals. 'Researcher
punches a LION in the face as it tries to EAT him in
Botswana: Male pal naked from the waist down saves him
by throwing elephant dung at it before another friend hits
it with car', www.dailymail.co.uk/news/article-9082071
/Researcher-punches-LION-face-tries-EAT-Botswana.
html

7 This story was covered by the *Standard*, a Kenyan
newspaper: www.standardmedia.co.ke/kenya/article/20012
50582/two-lions-kill-herder-in-nairobi-national-park

8 The death of Weldon Kirui, as covered by the UK newspaper
the *Evening Standard*, www.standard.co.uk/news/world/
teenager-mauled-to-death-by-lions-in-kenyan-national
-park-a3605946.html

9 The death of Kobus Marais, as reported in the South
African news outlet *SAPeople*, www.sapeople.com/2021
/02/08/dedicated-ranger-fatally-mauled-by-starving-lion
-whilst-on-rhino-patrol

10 Covered in Frank, L., Hemson, G., Kushnir, H., Packer, C.
& Maclennan, S. D. 2008. Lions, conflict and conservation.
Management and conservation of large carnivores in west
and central Africa 81–98 (reference available at http://
leofoundation.org/wp-content/uploads/2015/03/2008
-Management-and-conservation-of-large-carnivores-....pdf
#page=81)

11 A report, prepared for Sociedade para a Gestão e
Desenvolvimento da Reserva do Niassa Moçambique
by Colleen Begg *et al.* titled Preliminary data on
human–carnivore conflict in Niassa National Reserve,
Mozambique, particularly fatalities due to lion, spotted
hyaena and crocodile. Available at https://biofund.org.

mz/wp-content/uploads/2019/01/1548842722-F1183. Preliminary%20data%20on%20human%20-%20 carnivore%20conflict%20in%20Niassa%20National%20 Reserve.PDF

12 Data from Dunham, K. M., Ghiurghi, A., Cumbi, R. & Urbano, F. 2010. Human–wildlife conflict in Mozambique: a national perspective, with emphasis on wildlife attacks on humans. *Oryx* 44(2):185–93. Available at www.cambridge. org/core/journals/oryx/article/humanwildlife-conflict-in -mozambique-a-national-perspective-with-emphasis-on -wildlife-attacks-on-humans/434EEAAF88F3C10E9FA 6B55F2C3ACE39

13 Breaking the earlier rule that such attacks go largely unnoticed by international media, these attacks were covered in the *Irish Examiner* www.irishexaminer.com/ world/arid-30221673.html

14 Reported in *Addis Insight* www.addisinsight.net/lion-pride -attack-kills-people-in-ethiopia

15 Documented in Treves, A. & Naughton-Treves, L. 1999. Risk and opportunity for humans coexisting with large carnivores. *Journal of Human Evolution* 36(3):275–82. Available at www.sciencedirect.com/science/article/abs/pii /S0047248498902688

16 Covered by BBC at www.bbc.co.uk/news/world-africa -40656863

17 A narrow escape, documented at https://allafrica.com/ stories/202005140624.html

18 Covered briefly by UK newspaper the *Sun* at www.thesun. co.uk/archives/news/681935/man-eaten-in-zambia-lion -attack

19 Covered by the *Zambian Observer*, with the particularly prosaic headline 'Man eaten by a lion in Mfuwe', available at www.zambianobserver.com/man-eaten-by-a-lion-in -mfuwe

20 An account of the lion's predatory activities and eventual demise are given in www.chicagotribune.com/news/ct-xpm -1998-09-03-9809040001-story.html

21 Some idea of the extent of human predation by wildlife in Zambia can be got from: Chomba, C., Senzota, R., Chabwela, H., Mwitwa, J. & Nyirenda, V. 2012. Patterns of human wildlife conflicts in Zambia, causes, consequences and management responses. *Journal of Ecology and the Natural Environment* 4(12):303–13. Available at https://pdfs. semanticscholar.org/f500/6902b64e811cb65da78ab0702ad 047ac8baa.pdf

22 As stated in Bauer, H., de Iongh, H. & Sogbohossou, E. 2010. Assessment and mitigation of human–lion conflict in West and Central Africa. Available at www.degruyter.com/ document/doi/10.1515/mamm.2010.048/html

23 The increase in lions in India was covered by the *Times of India* in 2020 https://timesofindia.indiatimes.com/india/ roaring-success-population-of-asiatic-lions-in-india-up-29 -in-5-years/articleshow/76311768.cms

24 See this report for a summary www.downtoearth.org.in/ news/wildlife-biodiversity/92-lions-already-dead-in-gir -this-year-report-flags-71721

25 Kagathara, T. & Bharucha, E., 2020. Building walls around open wells prevent Asiatic Lion *Panthera leo persica* (Mammalia: Carnivora: Felidae) mortality in the Gir Lion Landscape, Gujarat, India. *Journal of Threatened Taxa* 12(3):15301–10. Available at http://threatenedtaxa.org/ index.php/JoTT/article/view/5025/6673

26 Lion deaths are covered in the article available at https:// science.thewire.in/environment/313-gir-lions-dead-in-two -years-where-do-govts-conservation-efforts-stand

27 The newspaper *India Today* covers lion deaths at www. indiatoday.in/india/story/222-lions-died-in-gir-forest -region-in-last-two-years-minister-1570238-2019-07-17

28 Covered in www.thehindu.com/news/national/virus -outdated-census-method-mar-gujarats-lion-numbers/ article31806558.ece

29 As reported here: www.downtoearth.org.in/news/wildlife -biodiversity/92-lions-already-dead-in-gir-this-year-report -flags-71721

30 Data from Jhala, Y. V., Banerjee, K., Chakrabarti, S., Basu, P., Singh, K., Dave, C. & Gogoi, K. 2019. Asiatic lion: Ecology, economics, and politics of conservation. *Frontiers in Ecology and Evolution* 7:312. Available at www.frontiersin.org/articles/10.3389/fevo.2019.00312/full

31 The Cecil the Lion story and resulting media outrage is thoroughly covered in Somerville, K. 2017. Cecil the lion in the British media: The pride and prejudice of the press. *Journal of African Media Studies* 9(3):471–85. Available at www.researchgate.net/profile/Keith-Somerville/publication /323658942_Cecil_the_lion_in_the_British_media_The_ pride_and_prejudice_of_the_press/links/60116e414585151 7ef1ab18b/Cecil-the-lion-in-the-British-media-The-pride -and-prejudice-of-the-press.pdf

32 Reflections on the Cecil story as covered in the Zimbabwean newspaper the *Herald*, available at www. herald.co.zw/reflections-on-our-cecil-the-zimbabwean -lion

33 See 31

34 More Zimbabwean reflections, this time covered by the BBC, available at www.bbc.co.uk/news/world-africa-3372 2688

35 See 31

36 The death of Xanda, as reported by the BBC, available at www.bbc.co.uk/news/world-africa-40671590

37 Voortrekker the elephant's death, reported by *Africa Geographic*, available at https://africageographic.com/ stories/iconic-desert-adapted-elephant-voortrekker-killed -by-trophy-hunter-in-namibia

38 The role of social media in public perception of large carnivores, especially the issue of misinformation in conservation, is investigated by Nanni, V., Caprio, E., Bombieri, G., Schiaparelli, S., Chiorri, C., Mammola, S., Pedrini, P. & Penteriani, V. 2020. Social media and large carnivores: Sharing biased news on attacks on humans. *Frontiers in Ecology and Evolution*. Available at www. frontiersin.org/articles/10.3389/fevo.2020.00071/full

39 Tooth and Claw has, to date, three series of four audio episodes each covering predators from army ants to tigers. Available via BBC World Service, you can find them by searching for 'Tooth and Claw' at www.bbc.co.uk/programmes/p002w557

40 Information taken from Kushnir, H. & Packer, C. 2019. Perceptions of risk from man-eating lions in southeastern Tanzania. *Frontiers in Ecology and Evolution* 7:47. Available at www.frontiersin.org/articles/10.3389/fevo.2019.00047/full

41 The risk calculator tool is available at www.qcovid.org/calculation

42 The lethal attack was covered by a number of media outlets, including *National Geographic*, available at httpnationalgeographic.com/animals/article/woman-mauled-death-lion-kevin-richardson-lion-whisperer-south-africa-private-reserves-lion-walks-spd

43 The role of fences (and money) is discussed in Packer, C., Loveridge, A., Canney, S., Caro, T., Garnett, S. T., Pfeifer, M., Zander, K. K., Swanson, A., MacNulty, D., Balme, G. & Bauer, H. 2013. Conserving large carnivores: dollars and fence. *Ecology Letters* 16(5):635–41. Available at https://onlinelibrary.wiley.com/doi/abs/10.1111/ele.12091

44 The counter-argument is given in Creel, S., Becker, M. S., Durant, S. M., M'soka, J., Matandiko, W., Dickman, A. J., Christianson, D., Dröge, E., Mweetwa, T., Pettorelli, N. & Rosenblatt, E. 2013. Conserving large populations of lions – the argument for fences has holes. *Ecology Letters* 16(11):1413–e3. Available at https://onlinelibrary.wiley.com/doi/full/10.1111/ele.12145

45 These abuses are covered here: www.theguardian.com/global-development/2020/feb/07/armed-ecoguards-funded-by-wwf-beat-up-congo-tribespeople

46 The positive role of *bomas* in keeping livestock safe is discussed at https://africanconservation.org/kenya-lion-proof-homes-to-save-king-of-the-jungle-from-herders-spears-2

47 Hazing effectiveness was studied by Petracca, L. S., Frair, J. L., Bastille-Rousseau, G., Hunt, J. E., Macdonald, D. W., Sibanda, L. & Loveridge, A. J. 2019. The effectiveness of hazing African lions as a conflict mitigation tool: implications for carnivore management. *Ecosphere* 10(12):e02967. Available at https://esajournals.onlinelibrary.wiley.com/doi/full/10.1002/ecs2.2967

48 The killing of the so-called Marsh Pride was reported by *National Geographic,* available at www.nationalgeographic.com/magazine/article/poisoning-africa-kenya-maasai-pesticides-lions-poachers-conservationists

49 The chemical immobilisation of wild animals, and the issues surrounding the process, are well covered in a manual titled *The Chemical Immobilization of Wild Animals,* available at http://cza.nic.in/uploads/documents/publications/english/Final%20Mannual%20Chemical%20Immobilization%20of%20Wild%20Animals%20(1).pdf

Chapter Three: Tigers

1 The largest big cats in the world, including lion-tiger hybrids, are discussed at www.guinnessworldrecords.com/world-records/largest-feline-carnivore#:~:text=In+captivity%2C+the+largest+tiger,423+kg+(932+lb)

2 There are many different sources for this weight, some of which use 316kg, but either way that's a big lion. https://helpfulhyena.com/what-is-the-biggest-lion-that-has-ever-been-recorded

3 You see photographs of these nine subspecies at www.livescience.com/29822-tiger-subspecies-images.html

4 The paper proposing two subspecies is Wilting, A., Courtiol, A., Christiansen, P., Niedballa, J., Scharf, A. K., Orlando, L., Balkenhol, N., Hofer, H., Kramer-Schadt, S., Fickel, J. & Kitchener, A. C. 2015. Planning tiger recovery: understanding intraspecific variation for effective conservation. *Science Advances* 1(5):e1400175. Available at www.ncbi.nlm.nih.gov/pmc/articles/PMC4640610

5 Details can be found in Liu, Y. C., Sun, X., Driscoll, C., Miquelle, D. G., Xu, X., Martelli, P., Uphyrkina, O., Smith, J. L., O'Brien, S. J. & Luo, S. J. 2018. Genome-wide evolutionary analysis of natural history and adaptation in the world's tigers. *Current Biology* 28(23):3840–9. Available at www.sciencedirect.com/science/article/pii/S0960982218312144

6 The importance of understanding subspecies for conservation is discussed in an article by Time Crowe and Paulette Bloomer in *The Conversation*, available at https://theconversation.com/a-new-approach-to-understanding-subspecies-can-boost-conservation-68364

7 This estimate, and other tiger facts, can be found via WWF (which seems to back the two-subspecies classification scheme in footnote 4 www.worldwildlife.org/species/tiger

8 Details at www.newscientist.com/article/2250138-endangered-tigers-have-made-a-remarkable-comeback-in-five-countries

9 Interview available via BBC World Service at www.bbc.co.uk/programmes/w3ct2g91

10 Data from Bhattarai, B. R. & Fischer, K. 2014. Human–tiger *Panthera tigris* conflict and its perception in Bardia National Park, Nepal. Oryx 48(4):522–8. Available at www.cambridge.org/core/journals/oryx/article/humantiger-panthera-tigris-conflict-and-its-perception-in-bardia-national-park-nepal/1E6A59CC709E40565145CA6D44B74A99

11 Narrative and details on these attacks is given at www.thethirdpole.net/en/nature/why-have-tiger-attacks-spiked-in-nepals-bardia-national-park and at https://the himalayantimes.com/nepal/tiger-kills-five-in-a-month-in-bardiya

12 All of Jim Corbett's books are worth reading in terms of understanding tiger and leopard predation in early-twentieth-century India. The titles are mentioned in the text and I have drawn from most of them in this section. The short book *Treetops* is where he recounts his experiences with Princess, and then Queen, Elizabeth

13 See first source in 11

14 See 11

15 Details of the community forestry programme can be found
    at www.worldcat.org/title/community-forestry-in-nepal-a
    -policy-innovation-for-local-livelihoods/oclc/690934004

16 Covered in the *Nepal Times* at www.nepalitimes.com/
    opinion/villagers-step-up-to-protect-nepals-tigers

17 Details covered in an article by Raheja, N. M. in *Millennium
    Post* at www.millenniumpost.in/why-tigers-become-man
    -eaters-53401

18 Details on tiger territory range and others at https://ielc.
    libguides.com/sdzg/factsheets/tiger/behavior

19 Data from Barlow, A. C., Ahmad, I. U. & Smith, J. L. 2013.
    Profiling tigers (*Panthera tigris*) to formulate management
    responses to human-killing in the Bangladesh Sundarbans.
    Wildlife Biology in Practice 9(2):30–9. Available at www.
    researchgate.net/publication/260158565_Profiling_tigers
    _Panthera_tigris_to_formulate_management_responses
    _to_human-killing_in_the_Bangladesh_Sundarbans

20 Figure from Jagrata Juba Shangha. 2003. Human-wildlife
    interactions in relation to the Sundarbans reserved forest
    of Bangladesh. *Sundarbans Biodiversity Project Report.*
    Available   at   http://103.48.18.141/library/wp-content/
    uploads/2018/11/007-Human-wildlife-interaction-study.
    pdf

21 'Blood honey', the problem of tigers and some of the
    economics of the Sundarbans are discussed in an article
    by Dipanjan Sinha In *Business Insider India* at www.
    businessinsider.in/business/news/blood-honey-is-a
    -booming-business-and-the-rising-demand-is-changing
    -the-lives-of-moulis-in-sundarbans/amp_articleshow
    /85967092.cms

22 Information and data from Debnath, A. 2020. Social
    rejection of tiger-widows of Sundarban, India. *Journal of
    Critical Reviews* 7(15):3174–9. Available at www.jcreview.
    com/admin/Uploads/Files/61dea7c97da630.06522819.pdf.
    Other information and background, as well as the chance to
    help, can be found at www.tigerwidows.org.

23 Data from Gurung, B., Smith, J. L. D., McDougal, C., Karki, J. B. and Barlow, A. 2008. Factors associated with human-killing tigers in Chitwan National Park, Nepal. *Biological Conservation* 141(12):3069–78. Available at www.sciencedirect.com/science/article/abs/pii/S0006320708003443

24 Md. Mahbubul Alam, Mohammad Abidur Rahman, Md. Khairul Islam, James Probert, Petra Lahann (2011). *Bangladesh Sundarbans Tiger Human Conflict Report: 2011* DOI:10.13140/RG.2.2.31418.24003

25 Interview on *BBC World Service* at www.bbc.co.uk/programmes/w3ct2g8z

26 Article by Rachel Nuwer at www.bbc.com/future/article/20191120-the-problem-of-indias-man-eating-tigers

Chapter Four: Crocodilians

1 Detailed in Haddad Jr, V. & Fonseca, W. C. 2011. A fatal attack on a child by a black caiman (*Melanosuchus niger*). *Wilderness & Environmental Medicine* 22(1):62–4. Available at www.wemjournal.org/article/S1080-6032(10)00374-1/pdf

2 More information on the Chinese alligator can be found at www.iucnredlist.org/species/867/3146005

3 More information on the American alligator can be found at www.iucnredlist.org/species/46583/3009637

4 Data from Langley, R. L. 2010. Adverse encounters with alligators in the United States: an update. *Wilderness & Environmental Medicine* 21(2):156–63. Available at www.wemjournal.org/article/S1080-6032(10)00066-9/fulltext

5 Data from Woodward, A. R., Leone, E. H., Dutton, H. J., Waller, J. E. & Hord, L. 2019. Characteristics of American alligator bites on people in Florida. *The Journal of Wildlife Management* 83(6):1437–53. Available at https://wildlife.onlinelibrary.wiley.com/doi/10.1002/jwmg.21719

6 Information and quotes from an article by Melissa Hogenboom at www.theindependentbd.com/arcprint/details/48476/2016-06-21

7 More details in the *Crocodile Specialist Newsletter* Volume 37, January to March 2018, published by IUCN Species

Survival Commission and available at www.iucncsg.org/365 _docs/attachments/protarea/9165a6d4d4d38607adc3b1e 8b9e93c5e.pdf

8 More details in the *Crocodile Specialist Newsletter* Volume 28, January to March 2009 published by IUCN Species Survival Commission and available at www. iucncsg.org/365_docs/attachments/protarea/CSG%20 -e8b500ff.pdf

9 See Chapter Three, footnote 25

10 Article available at www.bbc.co.uk/news/newsbeat-57399 864

11 For details see Leslie, A. J. & Spotila, J. R. 2000. Osmoregulation of the Nile crocodile, *Crocodylus niloticus*, in Lake St. Lucia, Kwazulu/Natal, South Africa. *Comparative Biochemistry and Physiology Part A: Molecular & Integrative Physiology* 126(3):351–65. Available at www.sciencedirect .com/science/article/pii/S1095643300002154

12 Data from a thesis by Alistair Graham, submitted in 1968 and available at http://ufdcimages.uflib.ufl.edu/aa/00 /00/75/90/00001/lakerudolfcrocodilesalistairgraham. pdf

13 An article on Gustave by Michael McRae published in 2005 in *National Geographic Adventure* provides most of the background and information here. The article is available at https://web.archive.org/web/20080917225801/ http://adventure.nationalgeographic.com/2005/03/gustave -crocodile/michael-mcrae-text/1

14 An excellent article in *The Conversation* by Simon Pooley gives good background and information on attacks, available at www.theconversation.com/when -and-where-do-nile-crocodiles-attack-heres-what-we -found-119037

15 Information from Pooley, S. 2016. The entangled relations of humans and Nile crocodiles in Africa, c. 1840–1992. *Environment and History* 22(3:421–54. Available at www. ingentaconnect.com/content/whp/eh/2016/00000022 /00000003/art00006

16 Report at www.lusakatimes.com/2019/03/29/crocodile
-attack-survivor-appeals-for-cropping-of-the-reptiles-in
-zambezi-river

17 Data and information from Pooley, S., Botha, H.,
Combrink, X. & Powell, G. 2020. Synthesizing Nile
crocodile *Crocodylus niloticus* attack data and historical
context to inform mitigation efforts in South Africa and
eSwatini (Swaziland). *Oryx* 54(5):629–38. Available at www.
cambridge.org/core/journals/oryx/article/synthesizing-nile
-crocodile-crocodylus-niloticus-attack-data-and-historical
-context-to-inform-mitigation-efforts-in-south-africa
-and-eswatini-swaziland/6398E2EA56FC62E2C518071
584658CD4

18 Data from Dunham, K. M., Ghiurghi, A., Cumbi, R. &
Urbano, F. 2010. Human–wildlife conflict in Mozambique:
a national perspective, with emphasis on wildlife attacks on
humans. *Oryx* 44(2):185–93. Available at www.cambridge.
org/core/journals/oryx/article/humanwildlife-conflict-in
-mozambique-a-national-perspective-with-emphasis-on
-wildlife-attacks-on-humans/434EEAAF88F3C10E9FA
6B55F2C3ACE39

19 Data taken from Scott, R. & Scott, H. 1994. Crocodile bites
and traditional beliefs in Korogwe District, Tanzania. *BMJ*
309(6970):1691–2. Available at www.ncbi.nlm.nih.gov/pmc
/articles/PMC2542670/pdf/bmj00471-0025.pdf

20 Pooley, S. 2016. A cultural herpetology of Nile crocodiles in
Africa. Conservation and Society 14(4):391–405. Available
at https://eprints.bbk.ac.uk/id/eprint/16398/3/16398.pdf

21 Adam Britton provides an accessible overview of the
situation with saltwater crocs in this article in *The
Conversation* https://theconversation.com/staying-safe-in
-crocodile-country-culling-isnt-the-answer-60252

22 For breathless coverage of a big croc eating a big shark you
can always trust a UK tabloid… www.dailymail.co.uk/news/
article-2718627/Brutus-giant-croc-pictured-eating-bull
-shark-star.html

23 A remarkable interaction detailed in Gallagher, A. J., Papastamatiou, Y. P. & Barnett, A., 2018. Apex predatory sharks and crocodiles simultaneously scavenge a whale carcass. *Journal of Ethology* 36(2):205–9. Available at https://link.springer.com/article/10.1007/s10164-018-0543-2

24 Quoted in https://phys.org/news/2021-06-reckoning-animal-preyliving-crocodile-country.html

25 Coverage in www.thesun.co.uk/news/15717784/worlds-deadliest-crocodile-attacks

26 Visit www.crocodile-attack.info to find out much more about crocodilian attacks.

27 Stories at www.bbc.com/news/world-australia-36376227; www.insider.com/australia-swimmer-survives-after-prying-crocodile-jaws-from-his-head-2021-1; www.abc.net.au/news/2020-09-25/saltwater-crocodile-bites-man-on-head-in-far-north-queensland-r/12701322; www.cbsnews.com/news/crocodile-attack-australia-teen-dared-swim-river-innisfail

28 Stories at https://thethaiger.com/hot-news/crocodile-attacks-and-kills-55-year-old-fisherman-in-indonesia and www.thesun.co.uk/news/11505794/crocodile-sliced-open-severed-head-indonesia

29 Data from Sideleau, B., Sitorus, T., Suryana, D. & Britton, A. 2021. Saltwater crocodile (*Crocodylus porosus*) attacks in East Nusa Tenggara, Indonesia. *Marine and Freshwater Research* 72(7): 978–86. Available at www.publish.csiro.au/MF/MF20237

30 Data from Sideleau, B. M., Edyvane, K. S. & Britton, A. R. 2016. An analysis of recent saltwater crocodile (*Crocodylus porosus*) attacks in Timor-Leste and consequences for management and conservation. *Marine and Freshwater Research*, 68(5):801–9 and Amarasinghe, A. T., Madawala, M. B., Karunarathna, D. S., Manolis, S. C., de Silva, A. & Sommerlad, R. 2015. Human-crocodile conflict and conservation implications of saltwater crocodiles

Crocodylus porosus (Reptilia: Crocodylia: Crocodylidae) in Sri Lanka. *Journal of Threatened Taxa* 7(5):7111–30.

31 Information from Patro, S. & Padhi, S. K. 2019. Saltwater crocodile and human conflict around Bhitarkanika National Park, India: a raising concern for determining conservation limits. *Ocean & Coastal Management* 182:104923. Available at www.sciencedirect.com/science/article/pii/S09645691193 00857

32 An account by Platt *et al.* appeared in *The Herpetological Bulletin* Number 75 2001, available at www.thebhs.org/publications/the-herpetological-bulletin/issue-number-75 -spring-2001/2764-hb075-04/file

33 A fascinating man – www.aldoleopold.org/post/bruce-wright

34 Detailed in an article by Anna Pointer in the *Times* at www.thetimes.co.uk/article/crocodile-massacre-of-troops -debunked-as-wartime-myth-k6vxhdfsq

35 Data from Manolis, S. C. & Webb, G. J. 2013, September. Assessment of saltwater crocodile (*Crocodylus porosus*) attacks in Australia (1971–2013): implications for management. In *Crocodiles Proceedings of the 22nd Working Meeting of the IUCN-SSC Crocodile Specialist Group*. Gland, Switzerland: IUCN 97–104. Available at www.researchgate. net/profile/Dinal-J-S-Samarasinghe/publication /272832281_Human-Crocodile_conflict_in_Nilwala_River _a_social_science_perspective/links/599d2f66aca272d ff12be12b/Human-Crocodile-conflict-in-Nilwala-River-a -social-science-perspective.pdf#page=98

36 Numbers taken from Caldicott, D. G., Croser, D., Manolis, C., Webb, G. & Britton, A. 2005. Crocodile attack in Australia: an analysis of its incidence and review of the pathology and management of crocodilian attacks in general. *Wilderness & Environmental Medicine* 16(3):143–59. Available at www.sciencedirect.com/science/article/pii/S10806032057 0380X

37 Taken from the Wildlife Trade Management Plan for the Saltwater Crocodile (*Crocodylus porosus*) in the Northern

Territory of Australia, 2016–20, published in 2015 by Northern Territory Government Department of Land Resource Management. Available at www.awe.gov.au/ sites/default/files/env/pages/e85c86ad-0af5-40ad-b07b -99a27eed17e6/files/nt-saltwater-crocodile-mgt-plan-2016 -20.pdf

38 See www.abc.net.au/news/rural/2020-11-10/hermes-mick -burns-plan-to-build-huge-crocodile-farm-nt/12823662

39 For details on saltwater crocodile sustainable use, see this factsheet on Saltwater Crocodile Harvest and Trade in Australia published by CITES https://cites.org/sites/ default/files/eng/prog/Livelihoods/case_studies/CITES _livelihoods_Fact_Sheet_2019_Australia_Crocodiles.pdf

40 Explored in this article www.independent.co.uk/news /world/australasia/australia-could-introduce-trophy -hunting-saltwater-crocodiles-within-year-10339819 .html

41 See https://qz.com/695669/australia-is-considering -crocodile-trophy-hunting-to-make-national-parks-safer -for-humans

42 For example, see www.abc.net.au/news/rural/2018-06-04/ katherine-river-crocodiles-prevent-locals-from-swimming /9764080 and www.thetimes.co.uk/article/australia -crocodile-attacks-fuel-calls-for-a-cull-vx668rdgv

43 To listen to the interview, visit www.bbc.co.uk/programmes /w3ct2g8y

44 Quote taken from www.theconversation.com/staying-safe -in-crocodile-country-culling-isnt-the-answer-60252

45 From an article in *The Conversation* at www.theconversation .com/crocodile-culls-wont-solve-crocodile-attacks-11203

46 Information from www.bbc.co.uk/news/world-asia-india -46983559

47 Crocodile farming in Kenya is explored in www.bbc.co.uk/ news/business-37218790

48 Details and images of crocodile farming in Myanmar can be found in this article in *Myanmar Times* www.mmtimes. com/news/upgrading-crocodile-farm.html

49 Data from Pagoda, L. R. 2017. Crocodile–human encounter patterns in Sri Lanka. *Prehospital and Disaster Medicine* 32(S1):S117. Available at www.cambridge.org/core/journals /prehospital-and-disaster-medicine/article/crocodile -human-encounter-patterns-in-sri-lanka/F9BE3925DB8 6EC78F5058B16FEF3F1D0

Chapter Five: Forest Legends

1 For the full text, visit www.classicshorts.com/stories/lvta. html

2 Some details of Du Chaillu (misspelt as Du Challiu) can be found in the *Richmond Times Dispatch* of July 1861, available at www.perseus.tufts.edu/hopper/text?doc=Perseus:text: 2006.05.0230:article=pos=109

3 Reported in a story by Richard Hackensacker in 2008 in Indian newspaper the *Tribune* at www.tribuneindia.com /2008/20080427/spectrum/nature.htm

4 The man's case is reported in Chianura, L. & Pozzi, F. 2010. Case Report: A 40-year-old man with ulcerated skin lesions caused by bites of safari ants. *American Journal of Tropical Medicine and Hygiene* 83(1):9. Available at www. ncbi.nlm.nih.gov/pmc/articles/PMC2912566

5 Taken from McGraw, W., Scott, C. C. & Shultz, S. Primate remains from African crowned eagle (*Stephanoaetus coronatus*) nests in Ivory Coast's Tai Forest: Implications for primate predation and early hominid taphonomy in South Africa. *American Journal of Physical Anthropology: The Official Publication of the American Association of Physical Anthropologists* 131,2(2006):151–65. Available at https:// onlinelibrary.wiley.com/doi/10.1002/ajpa.20420

6 Information from Naude, V. N., Smyth, L. K., Weideman, E. A., Krochuk, B. A. & Amar, A. 2019. Using web-sourced photography to explore the diet of a declining African raptor, the Martial Eagle (*Polemaetus bellicosus*). *The Condor: Ornithological Applications* 121(1):duy015; and Swatridge, C. J., Monadjem, A., Steyn, D. J., Batchelor, G. R. & Hardy, I. C. 2014. Factors affecting diet, habitat

selection and breeding success of the African Crowned Eagle *Stephanoaetus coronatus* in a fragmented landscape. *Ostrich* 85(1):47–55.

7 The events were witnessed by the artist D. M. Henry and are recorded in Steyn, P. 1983. Birds of Prey of Southern Africa: *Their Identification and Life Histories*. Croom Helm, Beckenham, UK.

8 More details of the attack are given in Hart, D. & Sussman, R. W. 2018. *Man the hunted: Primates, predators, and human evolution*. Routledge. I interviewed Donna Hart for a documentary on human predation for BBC Radio, available at www.bbc.co.uk/programmes/m000k9r5

9 The account can be read at www.africanraptors.org/simon -thomsett-on-the-african-crowned-eagle-part-2

10 Noted in Steyn, P. 1983. *Birds of Prey of Southern Africa: Their Identification and Life Histories*. Croom Helm, Beckenham, UK.

11 Reported in www.dailymail.co.uk/news/article-7457499/ Child-killed-EAGLE-pinned-viciously-pecked.html

12 The 2006 paper looking at skull damage is Berger, L. R. 2006. Brief communication: Predatory bird damage to the Taung type-skull of *Australopithecus africanus* Dart 1925. *American Journal of Physical Anthropology: The Official Publication of the American Association of Physical Anthropologists* 131(2):166–8. Available at www. semanticscholar.org/paper/Brief-communication%3A -predatory-bird-damage-to-the-Berger/15a0f813e5c4c97 8810bfee965fea1dcfdcb67f0

13 See Berger, L. R. & McGraw, W. S. 2007. Further evidence for eagle predation of, and feeding damage on, the Taung child. *South African Journal of Science* 103(11–12):496–8. Available at www.scielo.org.za/scielo.php?script=sci_arttext &pid=S0038-23532007000600013

14 More details in Scofield, R. P. & Ashwell, K. W. 2009. Rapid somatic expansion causes the brain to lag behind: the case of the brain and behavior of New Zealand's Haast's Eagle (*Harpagornis moorei*). *Journal of Vertebrate Paleontology*

29(3):637–49. Available at www.tandfonline.com/doi/abs
/10.1671/039.029.0325

15 Quote taken from article by Michael Casey in *ABC News* at
https://abcnews.go.com/Technology/extinct-zealand-eagle
-eaten-humans/story?id=8557686

16 Taken from an article in the *Independent* at www.
independent.co.uk/climate-change/news/maori-legend-of
-maneating-bird-is-true-1786867.html

17 Details from www.toxinology.com/fusebox.cfm?fuseaction
=main.snakes.display&id=SN0048

18 The demise of Grant Williams can be read about in a *New
York Times* report by David Herszenhorn published on
Thursday 10 October 1996. It was modified slightly for
a report that is available at www.anapsid.org/nyburm.
html

19 Reported in the *Daily Mirror* at www.mirror.co.uk/news/uk
-news/snake-owner-strangled-8ft-pet-11905317

20 The attack was written up as a brief note, published as
Branch, W. R. & Hacke, W. D. 1980. A fatal attack on
a young boy by an African rock python *Python sebae*.
Journal of Herpetology 14(3):305–7. Available at www.
jstor.org/stable/1563557

21 Recently I learnt a story of an adult man who had been eaten
by an African rock python in South Africa. It happened
several decades ago and involved someone who worked
at an explosives factory. In an echo of some of the big cat
attacks it appears the victim was defecating when the attack
happened. The snake killed and ate him. Subsequently, while
trying to cross a railroad track the snake got stuck, with the
bulge of the man trapped between the rails. The snake was
hit by a train, at which point its last meal became apparent.
I have been unable to verify the story further, but I have no
reason to doubt the honesty of the people who told me of
the incident.

22 Details in Fredriksson, G. M. 2005. Predation on sun bears by
reticulated python in East Kalimantan, Indonesian Borneo.
The Raffles Bulletin of Zoology 53(1):165–8. Available at

https://web.archive.org/web/20070811101110/http://rmbr. nus.edu.sg/rbz/biblio/53/53rbz165-168.pdf

23 One such report was by the BBC at www.bbc.co.uk/news/ world-asia-39427458

24 Find out more about the Aeta people at www.aetatribes.org

25 Study by Headland, T. N. & Greene, H. W. 2011. Hunter–gatherers and other primates as prey, predators, and competitors of snakes. *Proceedings of the National Academy of Sciences* 108(52):E1470–4. www.ncbi.nlm.nih.gov/pmc/ articles/PMC3248510

26 A great example of how you can sometimes find the information you need in unrelated studies. Height data taken from Allingham, R. R. 2008. Assessment of visual status of the Aeta, a hunter-gatherer population of the Philippines (an AOS thesis). *Transactions of the American Ophthalmological Society* 106:240. Available at www.ncbi. nlm.nih.gov/pmc/articles/PMC2646443

27 This, and many other well-researched facts on anacondas, available at www.livescience.com/53318-anaconda-facts. html

Chapter Six: Hyenas

1 For an excellent general book covering hyenas, including interactions with people, Keith Somerville's *Humans and Hyenas: Monster or Misunderstood* published in 2021 by Routledge Studies in Conservation and the Environment is well worth a look.

2 For more on the biomechanics of hyena skulls see Tanner, J. B., Dumont, E. R., Sakai, S. T., Lundrigan, B. L. & Holekamp, K. E. 2008. Of arcs and vaults: the biomechanics of bone-cracking in spotted hyenas (*Crocuta crocuta*). *Biological Journal of the Linnean Society* 95(2):246–55. Available at https://web.archive.org/web/20100615201256/www.people. umass.edu/jtanner/Tanner_et_al2008.pdf

3 For more on this, and hyenas in general, listen to this BBC World Service documentary I presented in 2022 – www. bbc.co.uk/programmes/w3ct3jzb

4 Details in Wingfield, J. C. 2006. Communicative behaviors, hormone–behavior interactions, and reproduction in vertebrates. In *Knobil and Neill's Physiology of Reproduction* (1995–2040). Academic Press.

5 For this information, and much more, visit https://hyena -project.com/hyenas

6 The classic work on this topic is Kruuk, H. 1972. The spotted hyena: a study of predation and social behaviour (No. Sirsi) a102104). Available at www.amazon.co.uk/Spotted-Hyena -Predation-Social-Behavior/dp/1626549052/ref=asc_df _1626549052

7 More in Abay, G. Y., Bauer, H., Gebrihiwot, K. & Deckers, J. 2011. Peri-urban spotted hyena (*Crocuta crocuta*) in Northern Ethiopia: diet, economic impact, and abundance. *European Journal of Wildlife Research* 57(4):759–65. Available at https://link.springer.com/article/10.1007/s103 44-010-0484-8

8 Based on Sonawane, C., Yirga, G. & Carter, N. H. 2021. Public health and economic benefits of spotted hyenas *Crocuta crocuta* in a peri-urban system. *Journal of Applied Ecology* 58(12):2892–902. Available at https://besjournals. onlinelibrary.wiley.com/doi/pdf/10.1111/1365-2664.14024

9 For more on this, I interviewed Sonawane (see 8) for this BBC World Service documentary www.bbc.co.uk/ programmes/w3ct3jzb

10 This fascinating study is Baynes-Rock, M. 2013. Local tolerance of hyena attacks in East Hararge Region, Ethiopia. Anthrozoös 26(3):421–33. Available at www.tandfonline. com/doi/abs/10.2752/175303713X13697429464438.

11. This paper is Fell, M. J., Ayalew, Y., McClenaghan, F. C. & McGurk, M. 2014. Facial injuries following hyena attack in rural eastern Ethiopia. *International Journal of Oral and Maxillofacial Surgery* 43(12):1459–64. Available at www.sciencedirect.com/science/article/abs/pii/S0901 502714002562

12 Reported by the BBC at www.bbc.co.uk/news/magazine -26294631

13 For more on the Musa Jelle attack see www.bbc.co.uk/news /world-africa-18931521

14 The attack is detailed at www.iol.co.za/the-star/news/ zimbabwean-boy-mauled-and-disfigured-by-hyena-on-the -way-to-recovery-after-surgery-b3f9e38d-e929-47b1-a1b0 -d494e5860257

15 Reconstructive surgery and background is covered at www.news24.com/you/news/local/look-i-have-two-eyes -prosthetic-nose-and-eye-for-boy-mauled-by-hyena -20210908

16 This attack is described at www.zimbabwesituation.com/ news/hyenas-attack-devour-chirumanzu-man

17 Covered by the *Guardian* at www.theguardian.com/world /2007/nov/21/kenya.xanrice

18 An interesting study, published as Yirga, G., De Iongh, H. H., Leirs, H., Gebrihiwot, K., Deckers, J. & Bauer, H. 2012. Adaptability of large carnivores to changing anthropogenic food sources: diet change of spotted hyena (*Crocuta crocuta*) during Christian fasting period in northern Ethiopia. *Journal of Animal Ecology* 81(5): 1052–5.

19 Details from Brain, C. K. 1983. *The Hunters or the Hunted?: An Introduction to African Cave Taphonomy*. University of Chicago Press.

20 This and more detail available in the excellent Kruuk, H. 2009. *Hunter and Hunted: Relationships Between Carnivores and People*. Cambridge University Press.

21 Data from Mills, M. G. L. 1990. *Kalahari Hyaenas: The Comparative Behavioral Ecology of Two Species*. Unwin Hyman, London.

22 This review of the brown hyena by Eaton was published in 1976 and is available at www.degruyter.com/document/doi /10.1515/mamm.1976.40.3.377/html

23 Documented in the *Indian Express* at https://indianexpress. com/article/cities/pune/viral-video-of-hyena-attack-70 -year-old-injured-by-hyena-was-warned-minutes-before -attack-by-videographer-7504679

24 From Heptner, V. G. ed., 1989. *Mammals of the Soviet Union, Volume 2 Part 2 Carnivora (Hyenas and Cats)*. Full text available at https://archive.org/stream/mammalsofsov221 992gept/mammalsofsov221992gept_djvu.txt

Chapter Seven: Other Cats

1 This attack, and others, are documented in a report by the Humane Society of the United States available at www. humanesociety.org/sites/default/files/docs/captive-big-cat -incidents.pdf

2 Reported at https://english.onlinekhabar.com/five-injured -in-saptari-leopard-attack.html. Note the photo of the leopard, but the text stating the animal was a clouded leopard.

3 Taken from www.theguardian.com/us-news/2018/may/20/ mountain-bikers-fatal-cougar-attack-washington-state

4 This remarkable account can be read at www.theguardian. com/lifeandstyle/2016/mar/11/i-fought-off-mountain-lion -experience

5 Covered at www.nbcnews.com/news/us-news/mountain -lion-sighting-prompts-lockdowns-two-california-schools -n1281808

6 This attack is documented at www.reuters.com/article/us -usa-cougar-idUSN2436680520080624

7 Reportsavailableatwww.cbc.ca/player/play/1888746051939; www.latimes.com/california/story/2021-08-28/wildlife -agents-kill-a-mountain-lion-after-it-mauls-a-five-year-old -boy; https://bc.ctvnews.ca/b-c-woman-attacked-by-cougar -on-her-own-property-airlifted-to-hospital-with-serious -injuries-1.5413840; and www.theguardian.com/us-news /2021/aug/29/california-mother-fights-off-mountain-lion -with-bare-hands-to-save-5-year-old-son

8 See Chapter Six, 20

9 For this quote, detail and notes on Baron's book *Beast in the Garden*, see www.vaildaily.com/news/one-small-tale-tells -the-larger-truth

10 Data from www.statista.com/statistics/206820/number-of
-visitors-to-national-park-service-sites-since-2010; www.
statista.com/statistics/254012/number-of-visitors-to-the
-great-smoky-mountains-national-park; www.statista.com
/statistics/191240/participants-in-hiking-in-the-us-since
-2006

11 Taken from www.outsideonline.com/outdoor-adventure/
exploration-survival/are-mountain-lion-attacks-common

12 Data and information from Gilbert, S. L., Sivy, K. J.,
Pozzanghera, C. B., DuBour, A., Overduijn, K., Smith, M.
M., Zhou, J., Little, J. M. & Prugh, L. R. 2017. Socioeconomic
benefits of large carnivore recolonization through reduced
wildlife-vehicle collisions. *Conservation Letters* 10(4):431–
9. Available at https://conbio.onlinelibrary.wiley.com/doi
/10.1111/conl.12280

13 For example see www.dailymail.co.uk/news/article-8832929
/Terrifying-moment-cougar-launches-hiker-Utah-stalking
-SIX-MINUTES.html and www.dailymail.co.uk/news/
article-9596199/Terrifying-moment-Utah-hiker-stalked
-menacing-mountain-lion-baring-teeth-him.html

14 For this and more information on jaguars see www.
livescience.com/27301-jaguars.html

15 Study published as Espinosa, S., Celis, G. & Branch, L. C.
2018. When roads appear jaguars decline: Increased access
to an Amazonian wilderness area reduces potential for
jaguar conservation. *PloS one* 13(1):e0189740. Available at
www.ncbi.nlm.nih.gov/pmc/articles/PMC5751993

16 The diet of jaguars is covered in Hayward, M. W., Kamler, J.
F., Montgomery, R. A., Newlove, A., Rostro-García, S., Sales,
L. P. & Van Valkenburgh, B. 2016. Prey preferences of the
jaguar *Panthera onca* reflect the post-Pleistocene demise
of large prey. *Frontiers in Ecology and Evolution* 3:148.
Available at www.frontiersin.org/articles/10.3389/fevo.2015.
00148/full

17 Feeding details taken from Guidelines for Captive
Management of Jaguars by Law *et al.* and published for
the Jaguar Species Survival Plan. The report is available at

https://web.archive.org/web/20120113131137/www.
jaguarssp.com/Animal%20Mgmt/JAGUAR%20
GUIDELINES.pdf

18 Details taken from Iserson, K. V. & Francis, A. M. 2015.
Jaguar attack on a child: case report and literature review.
Western Journal of Emergency Medicine 16(2):303. Available
at www.ncbi.nlm.nih.gov/pmc/articles/PMC4380383

19 Data from www.rcpch.ac.uk/resources/uk-who-growth
-charts-0-4-years

20 This attack is reported at www.kaieteurnewsonline.com
/2014/01/08/mother-of-puma-attack-victim-fears-for
-childrens-safety

21 Documented in Neto, M. F. C., Neto, D. G. & Haddad
Jr, V. 2011. Attacks by jaguars (*Panthera onca*) on humans
in central Brazil: report of three cases, with observation
of a death. *Wilderness & Environmental Medicine*
22(2):130–5. Available at www.wemjournal.org/article/
S1080-6032(11)00044-5/fulltext

22 Confirmed by De Paula, R. C., Campos Neto, M. F.
& Morato, R. G. 2008. First official record of human
killed by jaguar in Brazil. *Cat News* 49:14. Available at
https://ava.icmbio.gov.br/pluginfile.php/4592/mod_data
/content/15695/CENAP+Cat_news_Cunha_et_al_2008.
pdf

23 Explored in Miththapala, S., Seidensticker, J. & O'Brien,
S. J. 1996. Phylogeographic subspecies recognition in
leopards (*Panthera pardus*): molecular genetic variation.
Conservation Biology 10(4):1115–32. Available at https://
repository.si.edu/bitstream/handle/10088/4298/
Miththapala1996.pdf

24 Published as Uphyrkina, O., Johnson, W. E., Quigley, H.,
Miquelle, D., Marker, L., Bush, M. & O'Brien, S. J. 2001.
Phylogenetics, genome diversity and origin of modern
leopard, *Panthera pardus*. *Molecular Ecology* 10(11):2617–
33. Available at https://onlinelibrary.wiley.com/doi/abs/10.
1046/j.0962-1083.2001.01350.x

25 The full report is available at https://repository.si.edu
   /bitstream/handle/10088/32616/A_revised_Felidae_
   Taxonomy_CatNews.pdf

26 Information from Kissui, B. M. 2008. Livestock predation
   by lions, leopards, spotted hyenas, and their vulnerability
   to retaliatory killing in the Maasai steppe, Tanzania.
   Animal Conservation 11(5):422–32. Available at https://
   zslpublications.onlinelibrary.wiley.com/doi/abs/10.1111/j.
   1469-1795.2008.00199.x

27 Data from Wang, S. W. & Macdonald, D. W. 2006.
   Livestock predation by carnivores in Jigme Singye Wangchuck
   national park, Bhutan. *Biological Conservation* 129(4):558–
   65. Available at www.sciencedirect.com/science/article/abs/
   pii/S0006320705005197

28 Data from Stuart, C. T. 1986. The incidence of surplus killing
   by *Panthera pardus* and *Felis caracal* in Cape Province, South
   Africa. *Mammalia* (Paris) 50(4):556–8.

29 Reported in the *Times of India* at https://timesofindia.
   indiatimes.com/city/dehradun/leopard-kills-47-goats-in
   -nainitals-jalal-village/articleshow/76309176.cms

30 Data from www.ons.gov.uk/peoplepopulationandcommunity
   /personalandhouseholdfinances/incomeandwealth/bulletins
   /householddisposableincomeandinequality/financialyearen
   ding2019provisional

31 Corbett's book *The Maneating Leopard of Rudraprayag* details
   the attacks and the subsequent hunting of the leopard.

32 The effects of the flu pandemic on India are detailed at
   https://gulfnews.com/opinion/op-eds/how-the-spanish-flu
   -changed-the-course-of-indian-history-1.1584285312898

33 Data and information from https://indianexpress.com/
   article/india/the-man-eating-leopards-of-rudraprayag-jim
   -corbett-5259713

34 Data from https://statisticstimes.com/demographics/india/
   uttarakhand-population.php

35 For example, see www.standard.co.uk/news/london/woman
   -mauled-in-bed-by-fox-i-m-traumatised-and-fear-i-would
   -contract-rabies-a3868586.html

36 Data from https://cheetah.org/cheetah-2019/wp-content /uploads/2019/05/policy-for-human-leopard-conflict -management-in-india.pdf

37 This study is Athreya, V., Odden, M., Linnell, J. D., Krishnaswamy, J. & Karanth, U. 2013. Big cats in our backyards: persistence of large carnivores in a human dominated landscape in India. *PloS one* 8(3):e57872. Available at https://journals.plos.org/plosone/article?id=10. 1371/journal.pone.0057872

38 Information from Kumbhojkar, S., Yosef, R., Kosicki, J. Z., Kwiatkowska, P. K. & Tryjanowski, P. 2021. Dependence of the leopard *Panthera pardus fusca* in Jaipur, India, on domestic animals. *Oryx* 55(5):692–8. Available at www. cambridge.org/core/journals/oryx/article/dependence -of-the-leopard-panthera-pardus-fusca-in-jaipur-india -on-domestic-animals/BA42F9F1EA7E714EB933BDF A5DB68E40

39 Attacks detailed at www.independent.co.uk/news/world /africa/leopard-uganda-ate-child-elisha-nabugyere-son -ranger-toddler-kill-queen-elizabeth-national-park -a8339946.html; www.theguardian.com/world/2019/jun /06/toddler-killed-by-leopard-in-south-africas-kruger -park; www.dailymail.co.uk/news/article-7804911/Leopard -attacks-EATS-boy-five-savages-childs-four-year-old -friend.html; and www.news24.com/news24/leopard-kills -boy-hotel-guilty-20090330

Chapter Eight: Bears

1 This remarkable evolutionary tale is explored in Salesa, M. J., Antón, M., Peigné, S. & Morales, J. 2006. Evidence of a false thumb in a fossil carnivore clarifies the evolution of pandas. *Proceedings of the National Academy of Sciences* 103(2:379–82. Available at www.pnas.org/doi/10.1073/pnas. 0504899102

2 The bite force of pandas and other carnivores can be found in Christiansen, P. & Wroe, S. 2007. Bite forces and evolutionary adaptations to feeding ecology in carnivores.

Ecology 88(2):347–58. Available at www.academia.edu /239888

3 A study by Zhang, P., Wang, T., Xiong, J., Xue, F., Xu, H., Chen, J., Zhang, D., Fu, Z. & Jiang, B. 2014. Three cases giant panda attack on human at Beijing Zoo. *International Journal of Clinical and Experimental Medicine* 7(11):4515. Available at www.ncbi.nlm.nih.gov/pmc/articles/PMC42 76236

4 The Paddington Bear stories were written by Michael Bond. Paddington become famous first as a children's TV series but has now achieved movie-star status. He is a spectacled bear, fond of marmalade sandwiches in the stories.

5 Diet preferences taken from Peyton, B. 1980. Ecology, distribution, and food habits of spectacled bears, *Tremarctos ornatus*, in Peru. *Journal of Mammalogy* 61(4): 639–52.

6 Study by Jorgenson, J. P. & Sandoval-A, S. 2005. Andean bear management needs and interactions with humans in Colombia. *Ursus* 16(1):108–16. Available at https://in. art1lib.com/book/20982072/844add

7 Diet choices given in Wong, S. T. 2002. Food habits of Malayan sun bears in lowland tropical forest of Borneo. *Ursus* 13:127–36.

8 Study by Sethy, J. & Chauhan, N. S. 2013. Human-sun bears conflict in Mizoram, northeast India: impact and conservation management. *International Journal of Conservation Science* 4(3). Available at https://web.s. ebscohost.com/ehost/pdfviewer/pdfviewer?vid=0&sid =08b42a06-c5cf-4e90-b2f8-3aeb5e9dd90b%40redis

9 Taken from Singh, N., Sonone, S. & Dharaiya, N. 2018. Sloth bear attacks on humans in central India: implications for species conservation. *Human–Wildlife Interactions* 12(3):5. Available at https://digitalcommons.usu.edu/hwi/vol12/ iss3/5

10 Data from Rajpurohit, K. S. & Krausman, P. R. 2000. Human–sloth-bear conflicts in Madhya Pradesh, India. *Wildlife Society Bulletin* 393–9.

11　A technical and detailed study by Puckett, E. E., Etter, P. D., Johnson, E. A. & Eggert, L. S. 2015. Phylogeographic analyses of American black bears (*Ursus americanus*) suggest four glacial refugia and complex patterns of postglacial admixture. *Molecular Biology and Evolution* 32(9):2338–50. Available at https://academic.oup.com/mbe/article/32/9 /2338/1029474

12　Predatory prowess reported by Svoboda, N. J., Belant, J. L., Beyer, D. E., Duquette, J. F., Stricker, H. K. & Albright, C. A. 2011. American black bear predation of an adult white-tailed deer. *Ursus* 22(1):91–4. Available at https://bioone.org /journals/ursus/volume-22/issue-1/URSUS-D-10-00024.1 /American-black-bear-predation-of-an-adult-white-tailed -deer/10.2192/URSUS-D-10-00024.1.short

13　Report available at https://archive.org/details/canadianfie ldnat108otta/page/236/mode/2up

14　An extremely useful study by Herrero, S., Higgins, A., Cardoza, J. E., Hajduk, L. I. & Smith, T. S. 2011. Fatal attacks by American black bear on people: 1900–2009. *Journal of Wildlife Management* 75(3):596–603. Available at https://wildlife.onlinelibrary.wiley.com/doi/epdf/10. 1002/jwmg.72

15　Detail from Nabi, D. G., Tak, S. R., Kangoo, K. A. & Halwai, M. A. 2009. Comparison of injury pattern in victims of bear (*Ursus thibetanus*) and leopard (*Panthera pardus*) attacks. A study from a tertiary care center in Kashmir. *European Journal of Trauma and Emergency Surgery* 35(2):153–8. Available at https://link.springer.com/article/10.1007/ s00068-008-8085-x

16　Study by Shafaat Rashid, T. A. K., Gh Nabi, D. A. R. & Manzoor Ahmed Halawi, B. A. M. 2009. Injuries from bear (*Ursus thibetanus*) attacks in Kashmir. *Turkish Journal of Trauma & Emergency Surgery* 15(2):130–4. Available at https://jag.journalagent.com/travma/pdfs/UTD_15_2_130 _134.pdf

17　Information from www.adfg.alaska.gov/static/education/ wns/brown_bear.pdf

18 Taken from www.bearconservation.org.uk/brown-bear

19 Study by Lindqvist, C., Schuster, S. C., Sun, Y., Talbot, S. L., Qi, J., Ratan, A., Tomsho, L. P., Kasson, L., Zeyl, E., Aars, J. & Miller, W. 2010. Complete mitochondrial genome of a Pleistocene jawbone unveils the origin of polar bear. *Proceedings of the National Academy of Sciences* 107(11):5053–7. Available at www.ncbi.nlm.nih.gov/pmc/articles/PMC2841953

20 Study by Liu, S., Lorenzen, E. D., Fumagalli, M., Li, B., Harris, K., Xiong, Z., Zhou, L., Korneliussen, T. S., Somel, M., Babbitt, C. & Wray, G. 2014. Population genomics reveal recent speciation and rapid evolutionary adaptation in polar bears. *Cell* 157(4):785–94. Available at www.ncbi.nlm.nih.gov/pmc/articles/PMC4089990

21 Detail from Cahill, J. A., Heintzman, P. D., Harris, K., Teasdale, M. D., Kapp, J., Soares, A. E., Stirling, I., Bradley, D., Edwards, C. J., Graim, K. & Kisleika, A. A. 2018. Genomic evidence of widespread admixture from polar bears into brown bears during the last ice age. *Molecular Biology and Evolution* 35(5):1120–9. Available at https://academic.oup.com/mbe/article/35/5/1120/4844088

22 Reported in https://web.archive.org/web/20170331110208/www.science.smith.edu/msi/pdf/i0076-3519-439-01-0001.pdf

23 A useful study – Bombieri, G., Naves, J., Penteriani, V., Selva, N., Fernández-Gil, A., López-Bao, J. V., Ambarli, H., Bautista, C., Bespalova, T., Bobrov, V. & Bolshakov, V. 2019. Brown bear attacks on humans: a worldwide perspective. *Scientific Reports* 9(1):1–10. www.nature.com/articles/s41598-019-44341-w.pdf

24 I interviewed Bombieri for a BBC World Service documentary on bears, available at www.bbc.co.uk/programmes/w3ct2g90

25 Data from Smith, T. S. & Herrero, S. 2018. Human–bear conflict in Alaska: 1880–2015. *Wildlife Society Bulletin* 42(2):254–63. Available at https://wildlife.onlinelibrary.wiley.com/doi/abs/10.1002/wsb.870

26 A write-up of this study, plus links to the paper, can be found at https://polarbearscience.com/2017/07/12/polar -bear-attacks-are-extremely-rare-says-new-study-but-the -data-is-incomplete

27 Article at www.bbc.co.uk/news/magazine-33334431

28 Viewing etiquette available at www.nps.gov/subjects/bears/ viewing.htm

Chapter Nine: Canids

1 Data from Westgarth, C., Brooke, M. & Christley, R. M. 2018. How many people have been bitten by dogs? A cross-sectional survey of prevalence, incidence and factors associated with dog bites in a UK community. *J. Epidemiol Community Health* 72(4):331–6. Available at https://jech. bmj.com/content/72/4/331

2 For example, www.independent.co.uk/news/world/ americas/man-eaten-dogs-texas-freddie-mack-death -police-a9000036.html

3 Attack reported at https://metro.co.uk/2017/06/17/drunk -man-eaten-alive-by-pack-of-stray-dogs-6716387

4 Study by Woodroffe, R., Lindsey, P., Romanach, S., Stein, A. & ole Ranah, S. M. 2005. Livestock predation by endangered African wild dogs (*Lycaon pictus*) in northern Kenya. *Biological Conservation* 124(2):225–34. Available at www.sciencedirect .com/science/article/abs/pii/S0006320705000558

5 One of the many media reports – www.dailymail.co.uk/news /article-2227757/Pittsburgh-zoo-death-Maddox-Derkosh -mauled-death-African-wild-dogs-mother-dangles-railings .html

6 These attacks are documented at https://phys.org/news /2019-04-australia-urgent-spate-dingo.html

7 Study by Jenks, K. E., Songsasen, N., Kanchanasaka, B., Leimgruber, P. & Fuller, T. K. 2014. Local people's attitudes and perceptions of dholes (*Cuon alpinus*) around protected areas in southeastern Thailand. *Tropical Conservation Science* 7(4):765–80. Available at https://journals.sagepub. com/doi/full/10.1177/194008291400700413

8 Suggested by Johansson, M. & Karlsson, J. 2011. Subjective experience of fear and the cognitive interpretation of large carnivores. *Human Dimensions of Wildlife* 16(1):15–29. Available at www.tandfonline.com/doi/abs/10.1080/10871209.2011.535240

9 Available at www.timesofisrael.com/snacks-for-stray-cats-may-be-feeding-a-rash-of-jackal-attacks

10 Available at www.9news.com.au/world/india-news-jackal-attacks-nearly-40-people-in-five-villages/0e5282cb-3b54-4cfb-b6f3-0adade4a7098

11 Available at www.telegraphindia.com/west-bengal/jackals-kill-feed-on-9-year-old-boy-in-in-murshidabad/cid/1693120

12 Details from White, L. A. & Gehrt, S. D. 2009. Coyote attacks on humans in the United States and Canada. *Human Dimensions of Wildlife* 14(6):419–32. Available at www.tandfonline.com/doi/full/10.1080/10871200903055326

13 Study by Wilson, P. J. & Rutledge, L. Y. 2021. Considering Pleistocene North American wolves and coyotes in the eastern Canis origin story. *Ecology and Evolution* 11(13):9137–47. Available at https://onlinelibrary.wiley.com/doi/full/10.1002/ece3.7757

14 Report available at https://digitalcommons.unl.edu/cgi/viewcontent.cgi?article=1026&context=wolfrecovery

15 A good summary of the attacks is given at https://en.wikipedia.org/wiki/Kirov_wolf_attacks

16 See 14 for details of Pavlov's work and a summary of the Kirov wolf attacks

17 The report is available at www.adfg.alaska.gov/static/home/news/pdfs/wolfattackfatality.pdf

18 Reported at www.sasktoday.ca/north/local-arts/author-disputes-coroners-conclusion-in-death-of-kenton-carnegie-4823446

19 See 14

20 This work is detailed in an article in *The Conversation* at www.theconversation.com/how-to-live-with-large-predators-lessons-from-spanish-wolf-country-167326

Chapter Ten: Fish, Lizards and Primates

1 Shark-attack data from www.floridamuseum.ufl.edu/shark -attacks/yearly-worldwide-summary

2 Reported at www.telegraph.co.uk/news/worldnews/asia/ india/3163501/Mutant-fish-develops-a-taste-for-human -flesh-in-India.html

3 Watch this behaviour at www.youtube.com/watch?v =27rudJsMAdo

4 Taken from Haddad Jr, V. & Sazima, I. 2003. Piranha attacks on humans in southeast Brazil: epidemiology, natural history, and clinical treatment, with description of a bite outbreak. *Wilderness & Environmental Medicine* 14(4):249–54. Available at www.wemjournal.org/article/S1080-6032(03)70563-8/fulltext

5 Reported at http://noticias.terra.com.br/mundo/america -latina/homem-bebado-morre-apos-ser-atacado-por -piranhas-na-bolivia,583aff0dfbada310VgnCLD200000bbc ceb0aRCRD.html

6 Reported at www.independent.co.uk/news/world/americas /girl-6-dies-after-piranhas-eat-her-legs-when-canoe -capsizes-on-family-holiday-10022189.html

7 The shoaling behaviour of piranha is reported by Queiroz, H. & Magurran, A. E. 2005. Safety in numbers? Shoaling behaviour of the Amazonian red-bellied piranha. *Biology Letters* 1(2):155–7. Available at www.ncbi.nlm.nih.gov/pmc /articles/PMC1626212

8 For example, www.theguardian.com/world/2007/jun/04/1

9 For more details see www.historyextra.com/period/ victorian/cannibalism-at-sea-sailors-ate-the-cabin-boy

10 From www.scientificamerican.com/article/why-would-a -chimpanzee-at

11 From www.bbc.co.uk/news/science-environment-40405026

12 Reported in the *East African,* available at https://web. archive.org/web/20060619230813/www.primates.com/ chimps/drunk-n-disorderly.html

13 Reported at www.foxnews.com/science/chimps-killing -people-in-uganda

14 The report by Shadrack Kamenya is available at http:// mahale.main.jp/PAN/9_2/9(2)-06.html

Index